FLORA OF TROPICAL EAST AFRICA

GESNERIACEAE

Iain Darbyshire*†

Monocarpic, annual or perennial herbs, rarely shrubs, caulescent or acaulescent. Leaves opposite, pairs equal to unequal, more rarely alternate, sometimes unifoliate with the lamina of cotyledonary origin. Inflorescences axillary or rarely terminal, cymose, often laxly so, the flowers of each dichotomy paired, occasionally congested and capitate, rarely racemoid. Flowers nearly always hermaphrodite, often showy, more rarely cleistogamous with a reduced corolla. Calyx divided to the base into five sepals, or tubular and five-lobed, sometimes the three upper lobes only united. Corolla gamopetalous, tubular, lobes usually 5, imbricate, often arranged in a bilabiate limb. Stamens usually two or four, rarely five, adnate to the corolla tube; staminodes present or absent; anthers bithecous, opening by longitudinal slits, free or variously connate. Disk annular or cupular, often lobed or undulate, occasionally absent. Ovary superior, unilocular. Placentation parietal, each placenta bilamellate, these occasionally becoming fused centrally, the capsule then appearing biolocular; ovules numerous. Fruit a capsule, often linear, sometimes spirally twisted, or a fleshy berry. Seeds numerous, small, ellipsoid or fusiform, sometimes with hair-like appendages at each end; endosperm absent or very slight.

The African Violet family has some 140 genera (7 in mainland Africa) and approximately 2900 species, mainly in the tropics and subtropics but with a few genera in temperate Eurasia. They often favour shaded habitats and are frequently epiphytic or lithophytic. African genera belong to the subfamily *Cyrtandroideae* and either occur in Asia or have affinity with genera there. They are not closely related to the genera of the Neotropics; characters unique to that region are therefore omitted from the above family description.

1. Cymes condensed, capitate, immediately subtended by a
    cordiform bract-like leaf; capsule circumscissile ...   1. **Epithema** (p. 2)
   Cymes more open and lax, 1–many-flowered, not
    immediately subtended by a bract-like leaf; capsule
    dehiscing longitudinally 2

* The taxonomy of the east African Gesneriaceae presented here, particularly that of the genus *Streptocarpus*, owes much to the revisional work of Drs O.M. Hilliard and B.L. Burtt (E). Dr. Burtt is particularly thanked for his useful taxonomic advice on *Epithema*, *Streptocarpus* and *Schizoboea*. The family description is adapted from that of Hilliard & Burtt in F.Z. 8(3): 43 (1988).

† The provision of distribution data and herbarium collections by Stella Simiyu (EA), together with her useful criticism on an earlier draft of the manuscript, has greatly aided the completion of the treatment of *Saintpaulia*. Colin Watkins is also thanked for his input and for his enthusiasm regarding this difficult genus. Tamás Pócs is thanked for his communication on the population of *Saintpaulia* from Mafi Hill and he is thus accredited with co-authorship of the newly described *Saintpaulia ionantha* H.Wendl. subsp. *mafiensis* I.Darbysh. & Pócs.

© The Board of Trustees of the Royal Botanic Gardens, Kew, 2006

2. Corolla tube ≤ 3 mm long, shorter than the upper lip
    of the limb; calyx lobes longer than or rarely up to
    1 mm shorter than the corolla tube; anthers always
    exserted, conspicuous, usually bright yellow . . . . . .   4. **Saintpaulia** (p. 50)
    Corolla tube 4–40 mm long, longer than the upper
    lip of the limb; calyx lobes always shorter than the
    corolla tube; anthers usually included within the
    corolla tube, if exserted then not bright yellow 3
3. Caulescent or acaulescent herbs; if the former then
    inflorescences axillary (though sometimes appearing
    terminal due to suppression of the terminal
    vegetative bud); capsule spirally twisted at maturity or
    if almost straight then shortly cylindric, 5–17 mm
    long and either with alternate leaves or with one
    large principal leaf . . . .   2. **Streptocarpus** (p. 5)
    Caulescent herbs; inflorescences terminal (though
    often overtopped by lateral shoots); capsule
    straight, narrowly cylindric, 15–30 mm long; leaves
    opposite . . . .   3. **Schizoboea** (p. 48)

The assessment of threat made under the Conservation Notes for each species, subspecies or variety follows the categories and criteria of the IUCN (2001, IUCN Red List Categories and Criteria: Version 3.1).

1. **EPITHEMA**

Blume in Bijdr., Fl. Nederl. Ind. 14: 737 (1826); C.B. Clarke in Monogr. Phan. 5: 179 (1883); Baker & C.B. Clarke in F.T.A. 4(2): 501 (1906); B.L. Burtt in F.W.T.A. ed. 2, 2: 383 (1963) & in Fl. Cameroun 27: 5 (1984) & in Fl. Gabon 27: 5 (1985)

Weak, often sappy, caulescent herbs. Plants pubescent throughout, with some hairs characteristically hooked. Lowest leaf usually solitary, petiolate, flowering stems usually developing one pair of opposite, petiolate or subsessile leaves. Inflorescences appearing terminal but actually axillary to a bract-like cordiform leaf at the summit of the flowering stem, one to several such stems developing above the upper leaf pair; cymes dense, capitate, scorpioid, flowers paired. Calyx tubular towards the base, divided into five lobes approximately midway along its length. Corolla gamopetalous; tube longer than lobes; limb bilabiate, the upper lip suberect, two-lobed, the lower lip three-lobed. Stamens arising from near the corolla mouth, fused at the base, the posterior two only fertile, free section of filaments twisted, anthers cohering face to face; two anterior staminodes well-developed. Disk cylindric, enveloping the ovary, occasionally much reduced. Ovary unilocular, subglobose; style slender, held in position in the upper section of the corolla tube by the stamens; stigma extended beyond the stamens, papillose. Capsule thin-walled, subglobose, surrounded by the persistent, somewhat accrescent calyx, circumscissile. Seeds numerous, fusiform, ± striate.

An Old World genus of over 20 species (only one occurring in Africa), currently under revision by O.M. Hilliard & B.L. Burtt.

1. **Epithema tenue** *C.B.Clarke* in Monogr. Phan. 5: 181 (1883); Baker & C.B. Clarke in F.T.A. 4(2): 501 (1906); B.L. Burtt in F.W.T.A. ed. 2, 2: 383 (1963); F. Hallé & Delmotte in Adansonia, sér. 2, 13: 273–287 (1973); B.L. Burtt in Fl. Cameroun 27: 6, t. 1 (1984) & in Fl. Gabon 27: 7, t. 1 (1985). Type: Bioko, *Mann* 2345 (K!, holo.)

© The Board of Trustees of the Royal Botanic Gardens, Kew, 2006

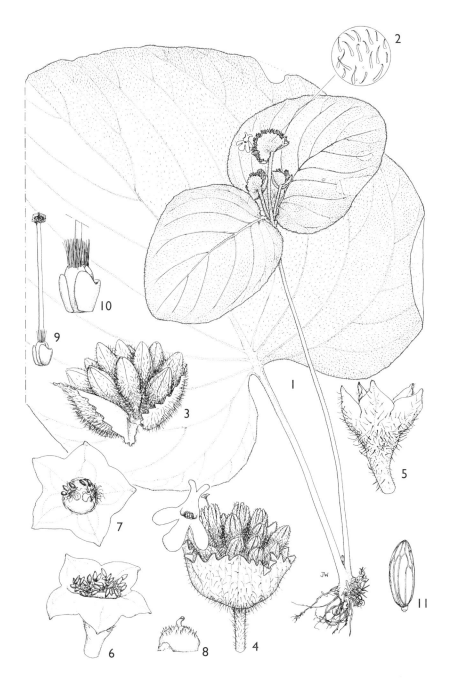

FIG. 1. *EPITHEMA TENUE* — **1**, habit × ²/₃; **2**, detail of leaf surface indumentum × 5; **3**, inflorescence with cleistogamous flowers × 3; **4**, inflorescence with mature chasmogamous flower and buds × 2; **5**, mature calyx × 3; **6**, fruiting calyx with seeds exposed due to loss of the capsule operculum × 6; **7**, fruiting calyx from above, showing attachment of seeds within capsule × 6; **8**, operculum of capsule × 6; **9**, pistil and disk × 6; **10**, detail of ovary within disk showing hirsute operculum × 12; **11**, seed × 40. 1, 2 from *Cable* 3256; 3 from photo taken by M. Cheek, Cameroon; 4 from *Etuge* 2057; 5, 11 from *Cable* 283; 6, 7, 8 from *A.S. Thomas* 1529; 9, 10 from *Cable* 19. Drawn by Juliet Williamson.

© The Board of Trustees of the Royal Botanic Gardens, Kew, 2006

Annual, sappy herb, 4–30 cm tall. Growth form variable, most commonly with a single large leaf on a stem to 12 cm long, above which develops a 1–14 cm long shoot terminating with a pair of smaller leaves, then 1–3 flowering shoots 0.7–5(–14) cm long, each with a reduced terminal leaf and subsessile axillary inflorescence; small plants may have only a single leaf; larger plants may develop multiple leaf-bearing shoots and/or more than one pair of opposite leaves per shoot. Stems sparsely pubescent with mainly hooked hairs. Lowest leaf blade broadly ovate to suborbicular, 6.5–21 cm long, 4–17 cm wide, base cordate, often deeply so, margin shallowly serrate to serrate-crenate, apex obtuse, surfaces pubescent, often sparsely so, hairs occasionally hooked, most dense at the margin; lateral nerves 5–13 pairs, ascending; petiole 1–5(–10) cm long, sparsely pubescent. Leaf pairs of flowering shoots subequal to unequal in size; blade 1.5–8(–14) cm long, 1–7(–14) cm wide, resembling lower leaf but base more commonly truncate to subcordate; petiole 0.2–3.5 cm long. Terminal leaf cordiform, enveloping the flowers on one side, 5–19 mm long, 12–24 mm wide, margin dentate, outer surface with short, often hooked hairs, inner surface largely glabrous. Inflorescences capitate, 5–24 mm broad; pedicels 0.5–1 mm long, extending somewhat in fruit, broad in one plane, puberulent with hooked hairs. Calyx pale green, campanulate, 3–7.5 mm long, tubular section sparsely pubescent and with short hooked hairs, lobes triangular-lanceolate, 1.5–3.5 mm long, ± densely long-pubescent outside giving the whole inflorescence a hairy appearance, glabrous within except for scattered stalked glands. Corolla of chasmogamous flowers pale blue or white, 8–13 mm long, sparsely pubescent or glabrous outside; tube 6–7 mm long, 2–2.5 mm diameter, broadening towards the open mouth; upper lip 4–4.5 mm long, divided in the upper half into two ovate lobes 1.5–2 mm long; lower lip divided almost to the base into three oblong lobes, 3.5–5.5 mm long, 1–2 mm wide; cleistogamous flowers much reduced, bud-like. Stamens with the free section of the filaments strongly recurved, ± 2 mm long; anther thecae rounded, ± 0.5 mm wide; staminodes well developed. Ovary 1.3–1.7 mm diameter, apically densely hirsute, the hairs long and straight, largely glabrous elsewhere; style to 7 mm long, glabrous; stigma capitate, 0.5–0.8 mm diameter. Capsule fragile, subglobose, surrounded by the persistent calyx, circumscissile. Seeds 0.4–0.5 mm long, longitudinally ridged, the ridges sometimes spiralling or convergent. Fig. 1, p. 3.

UGANDA. Bunyoro District: Budongo, May–June 1935, *Eggeling* 2010!; Toro District: Bwamba, Buranga, Sept. 1932, *A.S. Thomas* 729! & Bwamba, 1–2 km S of Sempaya, Sept. 1969, *Lye* 4286!
DISTR. **U** 2, 4; Guinea (Conakry) to Ivory Coast, eastern Nigeria to Gabon, Sudan and Uganda
HAB. Moist, mossy rocks or fallen tree trunks in wet forest; 300–1100 m
USES. None recorded on herbarium specimens
CONSERVATION NOTES. This species is widespread and locally common in suitable habitat, thus although local populations may be threatened by forest clearance, it is not currently threatened and is assessed as of Least Concern (LC)

NOTE. This species shares with *E. carnosum* Benth. from the Himalayas the feature of straight hairs on the operculum of the ovary, while the majority of species within the genus have hooked hairs (B.L. Burtt, *pers. comm.*). These two taxa are clearly closely allied, but differ in *E. carnosum* having shorter hairs on the operculum, in the inflorescence having a smaller subtending leaf and in the pedicels being more densely puberulent.

   Despite the significant number of herbarium specimens available of *E. tenue*, very few chasmogamous flowers are preserved, thus their description here is based largely on west African material, suitable flowers being unavailable in the east African specimens.

© The Board of Trustees of the Royal Botanic Gardens, Kew, 2006

2. STREPTOCARPUS

Lindl., Bot. Reg. 14, t. 1173 (1828); Baker & C.B. Clarke in F.T.A. 4(2): 504 (1906); B.L. Burtt in F.W.T.A. ed. 2, 2: 382 (1963); Hilliard & B.L. Burtt, Streptocarpus: An African Plant Study (1971) & in F.Z. 8(3): 43 (1988); Möller & Cronk in Syst. Geogr. Pl. 71: 545–555 (2001)

Linnaeopsis Engl. in E.J. 28: 483 (1900); Weigend in Flora 195: 49 (2000)

Monocarpic, annual or perennial herbs, rarely subshrubs; caulescent, then erect, creeping, or rarely subscandent, or acaulesent, then rosulate or unifoliate, sometimes with the 'leaf' of cotyledonary origin (phyllomorph). Leaves petiolate in the caulescent species, opposite-decussate, pairs subequal to unequal, rarely alternate or with multiple subsessile phyllomorphs alternating along an elongate stem-like axis with the lowest lamina disproportionately large; in acaulescent species the phyllomorph may show continued growth from a basal meristem, the lateral veins then widely spreading and parallel and the apex frequently withered due to seasonal drought stress; indumentum variable, inconspicuous sessile or subsessile glands usually present. Inflorescences axillary or, in some acaulescent species, from the "leaf-stalk" (***petiolode***) or base of the phyllomorph, rarely appearing terminal due to suppression of the terminal vegetative bud; cymose, often laxly branched, sometimes secund, rarely reduced to one or two flowers; bracts usually linear, inconspicuous, sometimes caducous; bracteoles as bracts but smaller. Calyx divided almost to the base into five lobes, rarely with a distinct tube (not in east Africa). Corolla gamopetalous, five-lobed, most often distinctly bilabiate, with a two-lobed upper lip and three-lobed lower lip, the latter often somewhat protruding with a central palate, the median lobe usually larger than the lateral pair. Stamens arising from within the corolla tube, the anterior two only fertile; two lateral staminodes usually present, often minute, the posterior staminode usually absent; filaments straight to convergent, sometimes recurved; anthers most often included within the corolla tube, more rarely exserted, thecae divaricate, rarely parallel, the two anthers connate at the apex, usually cohering face to face. Disk annular or shallowly cupular. Ovary unilocular but sometimes apparently bilocular by fusion of the T-shaped intrusive placentae; narrowed, often gradually, into the style; stigma variable, often with a central depression or bilobed, often papillose. Fruit a loculicidal capsule, cylindric, usually narrowly so, usually spirally twisting before maturity, rarely almost straight, dehiscence aided by slight untwisting in the former, when old sometimes splitting into four subvalves. Seeds numerous, small, fusiform. Seedlings with the cotyledons becoming unequal after germination.

A genus of over 130 species, confined to tropical and southern Africa and Madagascar. *Streptocarpus* is traditionally subdivided into two subgenera: *Streptocarpus*, comprising acaulescent unifoliate to rosulate taxa, and *Streptocarpella*, comprising caulescent, petiolate taxa with opposite phyllotaxy. However, although recent molecular data (e.g., Möller & Cronk in Syst. Geogr. Pl. 71: 545–555 (2001)) supports the recognition of two subgenera, it indicates that the dividing line is not so clear-cut. Subgen. *Streptocarpus* is found to include several caulescent taxa with opposite phyllotaxy, including *S. schliebenii* and *S. parensis* of east Africa. This agrees with Dr. Burtt's initial assertion that, on the basis of flower morphology, these two taxa were closely related to the acaulescent *S. montanus* (in Notes Roy. Bot. Gard. Edinb. 22: 576 (1958)), and demonstrates the significance of flower morphology in recognising affinities within the genus. However, at that time it led Burtt to the incorrect conclusion that *S. montanus* belonged within subgenus *Streptocarpella* on the basis that its rhizome represented a modified form of the typical caulescent habit within this group. This assertion was overturned following further morphological study which revealed the rhizome to be of hypocotyl origin rather than being a true stem (Hilliard & B.L. Burtt, Streptocarpus: 18–21 (1971)). However, the relationship between *S. montanus*, *S. parensis* and *S. schliebenii* was not then revisited.

© The Board of Trustees of the Royal Botanic Gardens, Kew, 2006

The genus *Linnaeopsis* Engl., comprising five taxa within three species, largely endemic to the Uluguru Mts of eastern Tanzania, is here placed in synonymy with *Streptocarpus*. *Linnaeopsis* was previously separated from *Streptocarpus* on the basis of the alternate phyllotaxy, the untwisted capsules and the small punctate pollen grains. Whilst the former character separates *Linnaeopsis* from most *Streptocarpus* species, it is not universally applicable. The subrosulate *L. alba* is very close in phyllotaxy to some of the rosulate members of subgenus *Streptocarpus*, and an incompletely known caulescent *Streptocarpus* from the Usambara Mts (sp. A here) apparently displays alternate leaves. As phyllotaxy is so variable in *Streptocarpus* therefore, the presence of alternate leaves seems of little taxonomic significance. Fruit morphologically is also likely of limited significance in generic definition, as not all *Streptocarpus* species have strongly twisted fruit (e.g. *S. exsertus* and *S. burttianus* in our region) and both *L. subscandens* and *L. alba* show some signs of slight twisting of the valves (though not to the same extent as seen in most *Streptocarpus*). However, the assertion by Weigend (in Flora 195(1): 45 (2000)) that *L. subscandens* has twisted fruits identical to that of some members of *Streptocarpus* may be in error; this was based upon the study of *Schlieben* 2936 (G) and material of the same collection from the Berlin herbarium contains a loose fruit which almost certainly belongs to a different species of *Streptocarpus*. In fact, in all the material seen of *Linnaeopsis*, very few mature fruits are available, thus our knowledge of their morphology remains incomplete. On the third point, although the small punctate pollen serves to confirm the affinity of the species within *Linnaeopsis*, it cannot be used to segregate this group from *Streptocarpus* where pollen form is diverse.

Most telling in the placement of *Linnaeopsis* is the strong similarity in flower morphology between the five taxa and several members of *Streptocarpus*, having small, white, obliquely subcampanulate corollas with stamens arising almost from the base of the tube. These characters are repeated in several species of subgen. *Streptocarpus* and almost identical to *S. bullatus*, a species also endemic to the Uluguru Mts. This affinity is confirmed by molecular evidence (Möller & Cronk in Syst. Geogr. Pl. 71: 545–555 (2001)) which clearly places *Linnaeopsis* with *S. bullatus* in one of the basal groups within subgen. *Streptocarpus*. This conclusion further highlights the importance of flower morphology in recognition of affinities within *Streptocarpus*.

Taxa within subgen. *Streptocarpus* often display unusual growth forms. Most commonly, during seed germination highly differential growth of the cotyledons occurs, with one developing rapidly into a large, leaf-like organ (phyllomorph) which may be held on a stalk of hypocotyl origin (petiolode). In certain species, the growth form becomes even more complicated, with vertical extension of the petiolode to form a stem-like structure (e.g. *S. bullatus*, *S. exsertus*) or fusion of multiple petiolodes into a rhizome-like structure (e.g. *S. montanus*, *S. burttianus*). For a full discussion on these growth forms, reference should be made to Hilliard & B.L. Burtt, Streptocarpus: 5–33 (1971). For the sake of simplicity, the cotyledon-derived phyllomorphs are referred to in the descriptions and key as leaves, but the term "petiolode" is adopted to separate these hypocotyl-derived leaf stalks, on which the inflorescence is often borne, from true petioles.

1. Caulescent herbs, leaves opposite or alternate 2
   Acaulescent herbs, sometimes with extended
      hypocotyl, leaves solitary to rosulate 24
2. Leaves opposite 3
   Leaves alternate, sometimes subrosulate 20
3. Leaf margin entire 4
   Leaf margin variously crenulate, serrulate,
      crenate or serrate 17
4. Palate of lower corolla lip not upcurved or
      ridged, mouth open or somewhat laterally
      compressed 5
   Palate of lower corolla lip upcurved and ±
      ridged, leaving mouth strongly vertically
      compressed or almost closed 12
5. Corolla tube < 10 mm long 6
   Corolla tube > 10 mm long 9

© The Board of Trustees of the Royal Botanic Gardens, Kew, 2006

6. Plants perennial, somewhat succulent, stem
    base often strongly swollen; seeds verru-
    culose; corolla 12–20 mm long .... **8. S. pallidiflorus** (p. 16)
    Plants annual, not succulent, stems sometimes
    fleshy but base not strongly swollen; seeds
    longitudinally ridged with scalariform
    transverse ridges; corolla 6–13.5 mm long 7
7. Corolla tube swollen centrally, narrowed
    towards the mouth .... **1. S. gonjaënsis** (p. 10)
    Corolla tube abruptly swollen in the upper
    half, widest at the mouth 8
8. Pedicels with scattered long glandular hairs;
    corolla 9–13.5 mm long, lower lip 3.5–6.5 mm
    long with widely spreading oblong lobes
    1.8–4 mm long .... **2. S. thysanotus** (p. 11)
    Pedicels with short eglandular hairs, rarely
    with occasional long glandular hairs;
    corolla 6.5–8.5 mm long, lower lip 2–3 mm
    long with less widely spreading, rounded
    lobes to 1.5 mm long .... **3. S. kimbozanus** (p. 12)
9. Leaf margin revolute; succulent plants of
    exposed rocky outcrops and cliffs 10
    Leaf margin not revolute; non-succulent
    plants of forest 11
10. Inflorescence 1–2-flowered; corolla ≤ 40 mm
    long; pedicels glabrous; leaves with 5–6
    pairs of lateral nerves .... **4. S. saxorum** (p. 13)
    Inflorescence at least 5-flowered; corolla ≥ 50
    mm long; pedicels glandular-pubescent;
    leaves with 9–12 pairs of lateral nerves ... **5. S. hirsutissimus** (p. 13)
11. Ovary glabrous; corolla ≤ 40 mm long ..... **14. S. euanthus** (p. 24)
    Ovary densely pubescent; corolla ≥ 40 mm
    long .... **15. S. bambuseti** (p. 25)
12. Ovary glabrous except for subsessile glands,
    sometimes with scattered hairs 13
    Ovary densely appressed-pubescent 16
13. Floor of corolla tube strongly ventricose in
    upper half, limb usually pale mauve with
    purple streaking on the lobes (particularly
    those of the upper lip), rarely darker and
    unstreaked .... **8. S. pallidiflorus** (p. 16)
    Floor of corolla tube not strongly ventricose,
    limb violet to blue-purple, often with a pale
    throat and palate, lobes unstreaked 14
14. Leaves densely pubescent on both surfaces,
    corolla tube ≤ 9 mm long .... **9. S. caulescens** (p. 18)
    Leaves sparsely pubescent (particularly when
    mature), mainly on the nerves of the lower
    surface, corolla tube ≥ 9 mm long 15
15. Corolla tube straight-cylindric, ≤ 12 mm long,
    not conspicuously inflated at the base ... **10. S. holstii** (p. 19)
    Corolla tube curved-cylindric, ≥ 14 mm long,
    conspicuously inflated above at the base .. **13. S. inflatus** (p. 24)

© The Board of Trustees of the Royal Botanic Gardens, Kew, 2006

© The Board of Trustees of the Royal Botanic Gardens, Kew, 2006

26. Corolla tube with long blunt unicellular hairs
 on the floor within; seeds reticulate; leaf
 margin entire . . . .   27. **S. rhodesianus** (p. 36)
 Corolla tube lacking long blunt unicellular
 hairs within; seeds verruculose; leaf margin
 dentate or crenate 27
27. Filaments distinctly U-curved (fig. 4.3);
 rhizome usually well-developed and bearing
 prominent leaf scars; corolla 11.5–19 mm
 long . . . .   19. **S. montanus** (p. 28)
 Filaments not U-curved; rhizome usually short
 and inconspicuous; corolla 6–12.5 mm
 long 28
28. Principal leaf 20–50(–100 cm) long, subsessile
 or on a short and inconspicuous petiolode
 to 2 cm long, lateral nerves spreading;
 primary peduncle 7–13 cm long . . . .   21. **S. burttianus** (p. 30)
 Principal leaf ≤ 21 cm long, borne on a
 slender stem-like petiolode 2.5–17 cm
 long, lateral nerves scalariform; primary
 peduncle 3–7 cm long . . . .   22. **S. bullatus** (p. 31)
29. Corolla to 14 mm long; lobes of lower lip
 oblong-spathulate; capsule 10–17 mm long,
 not or only slightly twisted . . . .   26. **S. exsertus** (p. 35)
 Corolla 20–60 mm long, lobes of lower lip
 rounded(-obovate); capsule (where known)
 20–80 mm long, strongly spirally twisted 30
30. Corolla tube straight to shallowly curved,
 cylindric for much of its length, not or only
 slightly deepened towards the mouth,
 2.3–3 times the length of the lower lip;
 indumentum of inflorescence always
 eglandular 31
 Corolla tube strongly curved and/or strongly
 deepened in the upper half to upper third;
 limb more conspicuous, the tube 0.8–2.4
 times the length of the lower lip;
 inflorescence with or without a glandular
 element to the indumentum 33
31. Corolla purple, 33–44 mm long; style 8–14 mm
 long; stigma 1.7–2 mm wide at its broadest
 point . . . .   30. **S. mbeyensis** (p. 39)
 Corolla white to pale lavender-blue, 22–35 mm
 long; style 3–8 mm long; stigma 0.9–1.2 mm
 wide at its broadest point 32
32. Stamens arising from the upper third of the
 corolla tube; pistil 15–24 mm long . . . . . .   28. **S. solenanthus** (p. 37)
 Stamens arising from the middle of the
 corolla tube; pistil ± 10 mm long
 (incompletely known species) . . . .   29. **S. sp. C** (p. 38)
33. Indumentum of inflorescence eglandular 34
 Indumentum of inflorescence with at least
 some glandular hairs on the pedicels,
 outside of the corolla and/or ovary, these
 often dense 35

© The Board of Trustees of the Royal Botanic Gardens, Kew, 2006

34. Corolla mouth laterally compressed, inverted-
    V-shaped; tube slender before abruptly
    deepening towards the mouth, 1.5–3 mm
    diameter in the lower half; capsule slender,
    1.5–1.7 mm diameter; calyx lobes 2–3.5 mm
    long (in East Africa) . . . .   31. **S. goetzei** (p. 40)
    Corolla mouth open, rounded; tube broader
    and more gradually deepened towards the
    mouth, (2.5–)3–4 mm deep in the lower
    half; capsule broader, 2–3 mm diameter;
    calyx lobes (3.3–)5–6.5 mm long . . . .   32. **S. kungwensis** (p. 41)
35. Anthers partially or wholly exserted beyond
    the corolla mouth; corolla tube ≥ 2 times as
    long as lower lip, the latter 9–11 mm long
    (incompletely known species) . . . .   35. **S. sp. D** (p. 46)
    Anthers included within the corolla tube, this
    ≤ 2 times as long as the lower lip, the latter
    9.5–28 mm long 36
36. Corolla mouth strongly laterally compressed
    and slit-like or inverted-V-shaped, the tube
    as long as or slightly shorter than the lower
    lip; ovary sparsely to densely glandular-
    pubescent or glabrous except for sessile
    glands . . . .   33. **S. compressus** (p. 42)
    Corolla mouth open and ± rounded, the tube
    1–2 times longer than the lower lip; ovary
    densely glandular and/or eglandular
    pubescent 37
37. Ovary usually with eglandular hairs only,
    more rarely with glandular hairs (perhaps
    of hybrid origin); corolla 24–35 cm long,
    with lilac to purple lobes within and
    lacking yellow markings at the mouth . . .   32. **S. kungwensis** (p. 41)
    Ovary with predominantly glandular hairs;
    corolla 37–60 mm long, or if < 35 mm
    (subsp. *chalensis*) long then with white
    lobes and yellow markings at the mouth .   34. **S. eylesii** (p. 43)

1. **Streptocarpus gonjaënsis** *Engl.* in E.J. 57: 210 (1921); B.L. Burtt in Notes Roy.
Bot. Gard. Edinb. 22: 575 (1958); Hilliard & B.L. Burtt, Streptocarpus: 340 (1971);
Iversen in Symb. Bot. Upsal. 28: 243 (1988). Types: Tanzania, Lushoto District, E
Usambara Mts, Gonja Mt, *Engler* 3355a (B†, holo., K!, illus.); Gonja Mt, above Mnyusi,
Peter 0 III 212 (B!, neo., selected by B.L.Burtt (1958), E!, photo.)

Annual caulescent herb, 10–25 cm tall. Stems usually unbranched, the lowest
5–10 cm being hypocotyl, sparsely spreading-pubescent. Lowest leaf pair of
cotyledonary origin, these highly unequal, the larger oblong-ovate, 5–9 cm long,
3–4 cm wide, base asymmetric, rounded, margin subentire, apex acute, sometimes
withered, the smaller 0.8–1.5 cm long; petiolode to 5 cm long; cauline leaves
opposite, subequal to somewhat unequal; blade ovate to oblong-ovate, 3–6 cm long,
1.5–2.5 cm wide, becoming smaller upwards, sparsely pubescent mainly on the upper
surface and nerves beneath; lateral nerves 8–10 pairs, ascending; petiole 0.5–1.8 cm
long. Inflorescences axillary and solitary or sometimes paired in the axil of the
principal cotyledonary leaf, 2–10-flowered; peduncles (1.5–)3–7.5 cm long, with
scattered glandular hairs; pedicels slender, 5–10 mm long, sparsely long glandular-

© The Board of Trustees of the Royal Botanic Gardens, Kew, 2006

pilose; bracts linear, 1–1.5 mm long, densely pilose. Calyx lobes narrowly lanceolate, 2–2.5 mm long, glandular- and eglandular-pilose. Corolla white, 6–6.5 mm long, with scattered glandular hairs outside; tube ± 5 mm long, centrally ventricose where 3 mm diameter, narrowed towards base and mouth, sparsely pilose inside below the stamens; limb bilabiate; upper lip of two erect lobes, 1–2 mm long and wide; lower lip protruding, of three broadly obovate lobes, 1.5–2.5 mm long and wide. Stamens arising in the middle of the corolla tube; filaments 2.5 mm long, glabrous; anther thecae rounded to subtriangular, 0.7 mm wide; staminodes minute. Ovary narrowly cylindric, 2–2.5 mm long, glabrous except for scattered subsessile glands; style 2–3 mm long, glabrous; stigma shallowly bilobed, 0.4–0.5 mm wide, papillose. Capsule narrow, 16–24 mm long, ± 1 mm diameter. Seeds 0.35–0.4 mm long, longitudinally ridged with scalariform transverse ridges.

TANZANIA. Lushoto District: Gonja Mt, above Mnyusi, Oct. 1915, *Peter* 0 III 196! & ibid., Nov. 1915, *Peter* 0 III 212! (type); Tanga District: Mt Mlinga, 1915, *Peter* 0 III 144!
DISTR. **T** 3; known only from the E Usambara Mts
HAB. Moist forest; 600–800 m
USES. None recorded on herbarium specimens
CONSERVATION NOTES. This poorly known species appears restricted to a narrow altitudinal range from two mountains in the east Usambara Mts, from where it was last collected in 1915. No direct attempts to subsequently rediscover this species at these sites have been made, thus its current status is uncertain, though with continued forest clearance at lower altitudes in this mountain range it is likely that it is threatened by habitat loss. It is therefore preliminarily assessed as Endangered (EN B1ab(iii)+2ab(iii))

NOTE. *S. gonjaënsis*, *S. thysanotus* and *S. kimbozanus* are closely allied to *S. elongatus* Engl. of Cameroon and São Tomé in west Africa, from which they differ in lacking the paired subsessile, broadly ovate or orbicular leaves at the stem apices characteristic of *S. elongatus*.

2. **Streptocarpus thysanotus** *Hilliard & B.L.Burtt* in Notes Roy. Bot. Gard. Edinb. 33: 467 (1975). Type: Tanzania, Morogoro District, Uluguru Mts, eastern slopes, seed coll. *Pócs*, cult. in R.B.G. Edinb., C.8094 (E!, holo.; EA!, K!, iso.)

Annual caulescent herb, 20–40 cm tall. Stems simple or branching in the lower half, fleshy, shortly eglandular-pubescent and with or without occasional longer glandular hairs; stomata sometimes conspicuous on young stems as purple streaks. Lowest leaf pair sometimes of cotyledonary origin, these highly unequal, the larger oblong-ovate, 8.5–10.5 cm long, 2.8–3.8 cm wide, base rounded, margin entire, apex acute, the smaller to 0.5 cm long; cauline leaves opposite, pairs subequal; blade sometimes purple-tinged beneath, ovate, 6–13.5 cm long, 2.5–8.5 cm wide, base asymmetric, cordate to obtuse, margin subentire, sometimes with marginal hydathodes visible, apex acute to shortly acuminate, surfaces sparsely pubescent, with or without occasional longer glandular hairs; lateral nerves 8–11 pairs; petiole 2–7.5 cm long, shortly pubescent, the hairs mainly eglandular. Inflorescences axillary, solitary, 4–30-flowered; peduncles slender, (2–)6–12.5 cm long, glabrous or with scattered long glandular hairs; pedicels wiry, 6–13 mm long, with scattered long glandular hairs; bracts linear, 1–3 mm long, pubescent, caducous. Calyx lobes lanceolate, 2–3 mm long, sparsely short eglandular-pubescent or glabrate. Corolla cream to pale purple, 9–13.5 mm long, glandular-pubescent outside; tube subcylindric, 5–7 mm long, 2.5–3 mm deep at the base, floor ventricose in the upper half, 3.5–4 mm deep at the open mouth, sparsely pilose inside below the stamens; limb bilabiate, highly oblique; upper lip of two erect lobes, (1–)1.5–2.5 mm long, 2–2.5 mm wide, pubescent inside; lower lip strongly protruding, 3.5–6.5 mm long, of three spreading oblong lobes 1.8–4 mm long, 1.4–2.2 mm wide, palate whitish with purple streaking. Stamens arising from the upper half of the corolla tube; filaments 2.5–3.5 mm long, verrucose towards the apex; anther thecae white or with purple spots, 0.8–1.5 mm wide; staminodes to 1.5(–2) mm long, pubescent. Ovary narrowly

© The Board of Trustees of the Royal Botanic Gardens, Kew, 2006

conical, 3.5–4.5 mm long, glabrous except for scattered subsessile glands; style 1.7–2.7 mm long, glabrous; stigma 0.4–0.8 mm wide, papillose. Capsule 17–36 mm long, 1–1.8 mm diameter, glabrous. Seeds 0.4–0.5 mm long, longitudinally ridged with scalariform transverse ridges.

TANZANIA. Morogoro District: Uluguru Mts, Mkungwe Forest Reserve, W and SE slopes above Kikundi village, Mar. 1987, *Pócs & Nsolomo* 87050/Z! & Mkungwe Forest Reserve, July 2000, *Mhoro* in UMBCP 36! & ibid., Oct. 2000, *Mhoro* in UMBCP 315!
DISTR. **T** 6; known only from the Uluguru Mts
HAB. Moist forest; 600–850 m
USES. None recorded on herbarium specimens
CONSERVATION NOTES. This species is currently known only from two sites in the Uluguru Mts, where it grows in low to mid altitude forest. It is susceptible to habitat loss either through human disturbance or through natural events such as forest fire. It is therefore here assessed as Vulnerable (VU D2), but may warrant elevation to Endangered status under criterion B if habitat degradation is found to be occurring within its known sites

NOTE. The cultivated specimen of *S. thysanotus* at R.B.G. Edinburgh differs from the wild material seen to date in having larger flowers which are pale purple, not cream, in colour.

3. **Streptocarpus kimbozanus** *B.L.Burtt* in Notes Roy. Bot. Gard. Edinb. 22: 575 (1958); Hilliard & B.L. Burtt, Streptocarpus: 340 (1971). Type: Tanzania, Morogoro District, Kimboza, *Padwa* 324 (E, holo.; EA, K!, iso.)

Annual caulescent herb, 20–40(–100) cm tall. Stems simple or branching in the lower half, fleshy, shortly eglandular-pubescent with occasional longer glandular hairs; stomata sometimes conspicuous on young stems as purple streaks. Leaves opposite, pairs subequal to unequal; blade oblong-ovate to oblong-elliptic, 6–7.5(–11.5) cm long, 2.5–3.5(–4.5) cm wide, base ± asymmetric, rounded to obtuse, margin subentire, sometimes with inconspicuous marginal hydathodes, apex acute, surfaces shortly pubescent; lateral nerves 10–12 pairs; petiole 2.5–4.5 cm long, shortly pubescent, hairs mainly eglandular. Inflorescences axillary, solitary, up to 10-flowered; peduncles slender, 2–3.5 cm long; pedicels wiry, 7–13 mm long, both sparsely short-pubescent, with or without very occasional long glandular hairs; bracts linear, to 1 mm long, pubescent, caducous. Calyx lobes lanceolate, 2–2.5 mm long, shortly eglandular-pubescent. Corolla white, 6.5–8.5 mm long, sparsely glandular-pubescent outside; tube cylindric, 4.5–5.5 mm long, 2 mm deep at the base, floor ventricose in the upper half, 2.8–3 mm deep at the open mouth, sparsely pilose inside below the stamens; limb bilabiate, highly oblique; upper lip of two erect lobes, 1–1.2 mm long and wide, shortly pubescent to glabrate inside; lower lip protruding, 2–3 mm long, of three rounded lobes 1.2–1.5 mm long, 1.3–1.6 mm wide, palate sometimes speckled purple at the throat. Stamens arising from the upper half of the corolla tube; filaments 1.5–2.5 mm long, verrucose towards the apex; anther thecae ± 0.75 mm wide; staminodes minute. Ovary narrowly conical, 2–2.8 mm long, glabrous except for scattered subsessile glands; style 2.3–3.5 mm long, glabrous; stigma ± 0.4 mm wide, papillose. Capsule 12–20 mm long, 1–1.5 mm diameter, glabrous. Seeds 0.45 mm long, longitudinally ridged with scalariform transverse ridges.

TANZANIA. Morogoro District: Uluguru Mts, Kimboza Forest Reserve, June 1983, *Polhill & Lovett* 4911! & ibid., Apr. 1986, *Pócs & Hall* 8653/L! & ibid., Sept. 2001, *Luke et al.* 7652!
DISTR. **T** 6; known only from the Uluguru Mts
HAB. Moist forest, particularly on damp rocks along streams; 200–550 m
USES. None recorded on herbarium specimens
CONSERVATION NOTES. This poorly known species is currently recorded only from a very narrow altitudinal range on the lower slopes of Mt Kimboza. Although this site is protected as a forest reserve, some degradation has occurred through past exploitation of timber, introduction of exotic timber trees (notably *Cedrela odorata* L.) and extraction of karstic deposits (W.R.Q. Luke, *pers. comm.*). This species is therefore assessed as Critically Endangered (CR B1ab(iii)+2ab(iii))

© The Board of Trustees of the Royal Botanic Gardens, Kew, 2006

NOTE. A scant specimen from along the Ruvu River (Jan. 1917, *Buchanan* s.n.!) appears to be of this species. The exact collecting location is unknown but may well refer to Kimboza Forest through which the Ruvu River flows.

4. **Streptocarpus saxorum** *Engl.* in E.J. 19: 154 (1895); Baker & C.B. Clarke in F.T.A. 4(2): 511 (1906); Hilliard & B.L. Burtt, Streptocarpus: 329 (1971); Hunt in Bot. Mag. 181: t. 740 (1977); B.L. Burtt in Fl. Pl. Afr. 46: t. 1807 (1980); Iversen in Symb. Bot. Upsal. 28: 237 (1988); Agnew, U.K.W.F. ed. 2: 264 (1994). Type: Tanzania, Lushoto District, Usambara Mts, Lutindi, *Holst* 3388 (B†, holo.; K!, iso.)

Succulent perennial herb. Basal stems straggling, woody, longitudinally wrinkled in dry material; flowering stems erect, densely pubescent or pilose; leaf scars prominent on bare lower stems. Leaves opposite, crowded in the upper section of the stems, pairs equal; blade fleshy, ovate-elliptic, 1.2–3(–3.8) cm long, 0.7–1.2(–2) cm wide, base obtuse to rounded, shortly attenuate in largest leaves, margin revolute, entire, apex obtuse, surfaces pubescent, densely so beneath particularly along the raised lateral nerves, some hairs glandular; lateral nerves 5(–6) pairs, ascending; petiole 0.2–1 cm long, pilose. Inflorescences axillary, solitary, 1–2-flowered; peduncles red-tinged, 7.5–13.5(–17) cm long, glabrate except for sparse glandular hairs towards base, peduncles of old inflorescences sometimes persisting at lower axils; pedicels 7–15 mm long, glabrous; bracts minute, pubescent. Calyx lobes oblong-lanceolate, 2–3.5 mm long, pubescent outside, with stalked glands within. Corolla white except for lobes pale- to mid-violet, (23–)27–40 mm long; tube 13–20 mm long, glandular-pilose outside, sparsely pilose to glabrous inside, saccate at the base, where ± 4.5 mm deep, narrowing to 2.5–4 mm before abruptly deepening to 4.5–6 mm midway along the tube, mouth open but laterally compressed; !limb bilabiate, highly oblique; upper lip reflexed, two lobes rounded, 5–10 mm long, 5.5–9.5 mm wide; lower lip protruding, (10–)13.5–23 mm long, lateral lobes 6–12 mm long, 6–13.5 mm wide, rounded, median lobe (6–)8–14 mm long, (6–)8–16 mm wide, sometimes slightly emarginate, shortly ciliate, palate sparsely pilose at the mouth. Stamens arising from immediately above the deepening of the corolla tube; filaments 1.7–2 mm long, pubescent with or without glandular hairs; anthers subtriangular, 1.7–2 mm wide, thecae indistinct; staminodes 1–2 mm long. Ovary narrowly conical, 6–8 mm long, densely appressed-pubescent and with stalked glands; style to 1 mm long, thick, down-curved; stigma flattened, shallowly bilobed, glutinous. Capsule 28–50 mm long, (2–)2.5–4.5 mm diameter, appressed-pubescent, style persistent. Seeds ± 0.45 mm long, longitudinally ridged with scalariform transverse ridges.

KENYA. Teita District: Teita Hills, Voi, Ndara Hill, undated, *Gardner* 3006! & Mt Kasigau, southern side above Bungule, Dec. 1970, *Faden* 70/972!; unlocated: *Glover* 2648!
TANZANIA. Tanga District: E Usambara Mts, Mlinga central peak, Oct. 1936, *Greenway* 4667!; Morogoro District: N Nguru Mts, Manyangu forest, N of Liwale R., Apr. 1953, *Drummond & Hemsley* 1987!; Iringa District: Mufindi, June 1971, *Paget-Wilkes* 914!
DISTR. **K** 7; **T** 3, 6, 7; not known elsewhere
HAB. On rocks and cliff faces, usually exposed to full sunlight, often in areas which receive periodic mist-derived moisture, occasionally near rivers; 500–2000 m
USES. Cultivated as an ornamental
CONSERVATION NOTES. Although somewhat localised in distribution, this species is recorded as locally common by several collectors within suitable habitat, where it may dominate the vegetation. Due to the inaccessibility and low agricultural potential of its rock and cliff habitats, there appears to be little anthropogenic threat. It is thus assessed as of Least Concern (LC)

5. **Streptocarpus hirsutissimus** *E.A.Bruce* in K.B.: 476 (1933); Hilliard & B.L. Burtt, Streptocarpus: 330 (1971); Lovett in Kew Mag. 6: 82–88 (1989). Type: Tanzania, Morogoro District, Uluguru Mts, Lupanga Peak, *B.D. Burtt* 3462 (K!, holo.)

© The Board of Trustees of the Royal Botanic Gardens, Kew, 2006

Succulent perennial herb. Basal stems woody, to 40 cm long, ± 1 cm diameter, sometimes straggling, longitudinally wrinkled in dry material, unbranched; upper stem erect, densely pale-villose particularly at the nodes, hairs multicellular; leaf scars prominent on bare lower stem. Leaves opposite, crowded in the upper section of the stems, pairs equal; blade elliptic, 3.5–6.5 cm long, 1.8–3 cm wide, apex acute, base rounded to attenuate, margin revolute, entire, densely villose above and on the raised lateral nerves and midrib beneath; lateral nerves 9–12 pairs; petiole 0.7–1.5(–2) cm long, densely villose. Inflorescences axillary, solitary, 5–10(–13)-flowered; peduncles 6–11 cm long, densely glandular-pubescent, peduncles of old inflorescences sometimes persisting at lower axils, then glabrescent; pedicels 7–20 mm long, densely glandular-pubescent; bracts to 3.5 mm long, often much shorter, densely villose. Calyx lobes oblong-lanceolate, 6–9 mm long, glandular- and eglandular-pubescent outside, becoming villose at the apex, sparsely puberulent within. Corolla pale blue-violet, paler at the mouth and somewhat darker on the tube, 50–70 mm, outside of tube and mouth glandular-pubescent; tube 30–40 mm long, basally saccate, where 6–10 mm deep, narrowing to 4.5–7 mm before deepening to 6.5–10 mm in the upper third, mouth open; limb bilabiate, highly oblique; upper lip reflexed, two lobes rounded, each 7–8.5 mm long, 7–10 mm wide; lower lip protruding, ± 30 mm long, lateral lobes rounded, 11–14 mm long, 14–17 mm wide, median lobe rounded to obovate, 14–18 mm long, 15–17 mm wide. Stamens arising in the upper third of the corolla tube; filaments 3.5–4(–5) mm long, sparsely glandular-pubescent; anthers suborbicular, 2.5–3.5 mm wide, thecae indistinct. Ovary narrowly cylindrical, 17–20 mm long, densely glandular-pubescent; style 10–14 mm long, glandular-pubescent; stigma bilobed, 1.5 mm wide. Capsule 55–85(–110) mm long, 2–3 mm diameter, glandular-pubescent. Seeds 0.6–1 mm long, longitudinally ridged with scalariform transverse ridges.

Tanzania. Morogoro District: Uluguru Mts, Lupanga Peak, May 1933, *B.D. Burtt* 4707! & N Nguru Mts, Kanga Mt, Dec. 1987, *Lovett & D.W. Thomas* 2636! & Uluguru Northern Catchment Forest Reserve, E side of ridge from Bondwa Peak to Nziwane, Jan. 2001, *Jannerup & Mhoro* 234!
Distr. **T** 6; known only from the Uluguru and Nguru Mts
Hab. Rock faces, cliffs and rocky montane grassland; 1650–2000 m
Uses. None recorded on herbarium specimens
Conservation notes. Known only from three locations, one of which is isolated, the future of this species is threatened by losses of populations through stochastic events such as rock falls or lightning-induced fires. It is therefore considered vulnerable (VU D2). It is, however, not greatly threatened by human influence, the populations being in rather inaccessible locations with little or no agricultural potential

6. **Streptocarpus stomandrus** *B.L.Burtt* in Notes Roy. Bot. Gard. Edinb. 22: 577 (1958); Hilliard & B.L. Burtt, Streptocarpus: 327 (1971); B.L. Burtt in Bot. Mag. 180: t. 690 (1975). Type: Tanzania, Morogoro District, Nguru Mts, fl. in hort. Amani, *Moreau* 900 (E, holo.; K!, iso.)

Suberect or decumbent caulescent herb, 20–25 cm tall. Stem bases succulent when young, becoming woody, branching, erect stems both long- and short-pubescent. Leaves opposite, pairs somewhat unequal to subequal; blade ovate-elliptic, 3.5–7 cm long, 2–4 cm wide, base rounded to shortly attenuate, margin shallowly crenulate or crenulate-serrulate, apex subacuminate to rounded, surfaces densely pubescent; lateral nerves (5–)7 pairs; petiole 0.5–2(–3.5) cm long, densely pilose. Inflorescences axillary to the upper leaves, solitary, lax, 4–16-flowered; peduncles (2–)5.5–11 cm long, pedicels 7–17 mm long, both spreading-pubescent with eglandular and some longer glandular hairs; bracts lanceolate, to 1 mm long, pubescent. Calyx lobes (oblong-)lanceolate, 1.5–3 mm long, densely pubescent. Corolla mauve, 25–32 mm long, sparsely glandular-pubescent outside; tube 12–14 mm long, slightly arcuate, expanded at the top and base, narrowing to 1.7–2.5 mm diameter in the centre,

© The Board of Trustees of the Royal Botanic Gardens, Kew, 2006

mouth open; limb bilabiate, oblique; upper lip of two erect lobes 2.5–4 mm long, 3.5–5 mm wide; lower lip 13–18 mm long, lobes spreading, lateral lobes 5.5–7 mm long, 6–8 mm wide, median lobe strongly protruding, rounded, 6–10 mm long, 6–9 mm wide, palate pale mauve with 2–4 purple markings centrally. Stamens arising from just within the corolla mouth; filaments 3.5–4.5 mm long, verrucose towards the apex; anthers exserted, subtriangular, 1.5–2 mm wide, thecae indistinct; staminodes minute. Ovary cylindric, 5.5–9 mm long, appressed-puberulent and/or with stalked glands, narrowing into the 3–9 mm long style, sparsely puberulent to glabrescent; stigma shallowly bilobed, 1.25–1.75 mm wide, papillose. Capsule 35–40 mm long, ± 1.5 mm diameter, puberulent to glabrescent, style persistent. Seeds ± 0.65 mm long, longitudinally ridged with scalariform transverse ridges.

TANZANIA. Morogoro District: Nguru Mts, Kombola, Aug. 1971, *Schlieben* 12229! & path from Magunga to Maskati, Mar. 1988, *Bidgood et al.* 436! & Kanga Forest Reserve, June 2005, *Haston* 66!
DISTR. **T** 6; known only from the Nguru Mts
HAB. Forest and amongst rocks in forest streams; 1000–1900 m
USES. Cultivated as an ornamental
CONSERVATION NOTES. This species has a highly limited distribution, currently known only from a restricted altitudinal range within the Nguru Mts. No data on abundance are recorded, but the paucity of collections (eight known) suggests that it is scarce. One specimen (*Bidgood et al.* 479) was collected in a "small forest patch" which may be taken to indicate that there has been some forest fragmentation within its range. This taxon is therefore likely to qualify as Endangered (EN B1ab(iii)+2ab(iii))

7. **Streptocarpus kirkii** *Hook.f.* in Bot. Mag. 110: t. 6782 (1884); Baker & C.B. Clarke in F.T.A. 4(2): 509 (1906); B.L. Burtt in K.B.: 84 (1939); Hilliard & B.L. Burtt, Streptocarpus: 328 (1971). Type: Tanzania, "hilly country of the coast opposite Zanzibar", *Kirk* s.n., fl. in hort., R.B.G. Kew (K!, holo.)

Caulescent herb, 10–45 cm tall. Stems woody and longitudinally wrinkled towards the base, where often decumbent, occasionally branching; upper stems fleshy, densely pubescent. Leaves opposite, pairs subequal; blade variable, sometimes purple-tinged beneath, elliptic to obovate, 1.8–8.5 cm long, 1–4 cm wide, base rounded, subcordate or attenuate, ± asymmetric, margin shallowly crenulate or crenulate-serrulate, apex obtuse-rounded to subacuminate, surfaces shortly pubescent; lateral nerves 6–7(–9) pairs; petiole 1–2(–3) cm long, shortly pubescent. Inflorescences axillary to the upper leaves, solitary, lax, up to 10-flowered; peduncles reddish, 4.5–9 cm long, extending to 11.5 cm in fruit, shortly pubescent, peduncles of old inflorescences often persisting at lower axils, then glabrescent; pedicels reddish, 14–20 mm long, pubescent and with scattered stalked glands; bracts linear, 1–2 mm long, pubescent. Calyx lobes lanceolate, 1.5–3 mm long, pubescent to glabrous except for scattered stalked glands. Corolla subcampanulate, somewhat declinate, pale- to mid-lilac, paler within the tube, sometimes with darker markings in the throat, (9.5–)12–18 mm long, pubescent outside; tube (5.3–)6.5–10 mm long, 2 mm diameter at the base, expanding to 4.5 mm at the open mouth where slightly upcurved, sparsely pubescent inside; limb directed forward, bilabiate, shortly ciliate; upper lip with lobes 2–2.5(–3) mm long, 2.5(–3) mm wide, rounded; lower lip protruding, (4.2–)5–8.5 mm long, lateral lobes 3–4 mm long, 3.7–4.7 mm wide, median lobe 3.5–5 mm long, 3.5–6 mm wide, all rounded. Stamens arising from the upper third of the corolla tube; filaments 2–2.5 mm long, curved, glabrous; anthers subtriangular, 1–1.3 mm wide, thecae indistinct; staminodes minute. Ovary cylindric, 3.5–6.5 mm long, with stalked glands, ± puberulent; style 2–4.5 mm long; stigma bilobed, 0.5–0.8 mm wide. Capsule 28–43 mm long, 1.5–2.5 mm diameter, glabrescent. Seeds 0.5–0.7 mm long, longitudinally ridged with scalariform transverse ridges.

© The Board of Trustees of the Royal Botanic Gardens, Kew, 2006

Kenya. Teita District: Sagala Hills near Voi, Jan. 1971, *Faden* 71/38A! & Taita Hills, Mbololo
Forest, May 1985, *Taita Hills Expedition* 360! & ibid., Dec. 2000, *Bytebier* 1938!
Tanzania. Lushoto District: Kiumba, Amani, Jan. 1950, *Verdcourt* 52!; Handeni District: Nyogi
Mt, Sept. 1933, *B.D. Burtt* 4869!; Morogoro District: Uluguru Mts, Nov. 1932, *Schlieben* 2940!
Distr. **K** 7; **T** 3, 6, 7; not known elsewhere
Hab. Usually epiphytic in forest, occasionally terrestrial; 900–1900 m
Uses. Cultivated as an ornamental
Conservation notes. This species has a rather wide range and has been noted as fairly
common in the Sagala Hills, Kenya (*Faden* 71/38A) and at Kwamkoro Forest Reserve,
Tanzania (*Semsei* 3541). It is likely overlooked due to its predominantly epiphytic habit. A
preliminary assessment of Least Concern (LC) is made here but further data on its current
distribution, abundance and threats to its habitat are required

Note. Inflorescence indumentum in this taxon is somewhat variable, ranging from rather
densely short-pubescent and with stalked glands in *Taita Hills Expedition* 360 from Kenya, to
largely glabrous in the type collection where the ovary has only stalked glands. Intermediate
degrees of indumentum are however observed in the additional material viewed.
   The origin of the type specimen is somewhat confused; on the herbarium sheet it states
"Zanzibar", but J.D. Hooker notes in the protologue that the plant grew from soil surrounding
ferns shipped back to R.B.G. Kew which were collected on the mainland in the coastal hills.

8. **Streptocarpus pallidiflorus** *C.B.Clarke* in F.T.A. 4(2): 508 (1906); Hilliard & B.L.
Burtt in Notes Roy. Bot. Gard. Edinb. 43: 231 (1986). Type: Tanzania, Kilimanjaro,
Marangu, *Volkens* 1006 (B†, holo.; K, lecto., selected by Hilliard & B.L. Burtt (1986),
missing; BM!, isolecto.)

Upright or sometimes straggling caulescent herb, 10–30(–60) cm tall. Stems
succulent, reddish, often strongly swollen at the base where up to 2 cm diameter, often
branching above the base, spreading-pubescent particularly when young. Leaves
opposite, pairs subequal; blade somewhat fleshy, ovate-elliptic, 1.5–6.5 cm long,
0.8–4 cm wide, base asymmetrically rounded or subcordate to shortly attenuate,
margin entire, apex acute to subacuminate, surfaces densely pubescent particularly
on the nerves beneath, rarely more sparsely so; lateral nerves 7–11 pairs; petiole
1–2(–3.5) cm long, densely pilose. Inflorescence axillary, solitary, 3–10-flowered;
peduncle 3–10(–15) cm long, pubescent towards the base, largely glabrous towards
the apex; pedicels fine, 10–20 mm long, glabrous or with scattered eglandular and/or
glandular hairs, rarely pubescent; bracts linear, 1–2 mm long, pubescent, often early
caducous. Calyx lobes purplish, ovate-lanceolate, 2–3 mm long, sparsely pubescent,
occasionally with glandular hairs, with or without scattered subsessile glands. Corolla
pale mauve or whitish, rarely purple, usually striped with dark pink to purple lines,
12–20 mm long; tube 6–9 mm long, strongly ventricose on the floor above the middle
where 3–4 mm deep, glandular-pilose outside, papillose within below the origin of the
stamens and at the mouth; limb bilabiate, oblique; upper lip of two erect, rounded
lobes 2–3 mm long and wide, lower lip 6–11 mm long, protruding, the three lobes
spreading, usually oblong-rounded, 3–6 mm long, 3–4.5(–6) mm wide, palate curved
upwards, weakly ridged so that the mouth is open but vertically compressed. Stamens
arising from the upper half of the corolla tube; filaments 2–3 mm long, straight,
glabrous; anther thecae divergent, ± 0.8 mm wide; staminodes minute. Ovary
cylindric, 4.5–6 mm long, glabrous except for scattered subsessile glands, rarely
sparsely pubescent; style 1.5–3 mm long, glabrous; stigma capitate, ± 0.4 mm wide,
papillose. Capsule 20–45 mm long, 1.5–2 mm diameter, glabrous. Seeds 0.4–0.6 mm
long, verruculose with slight longitudinal ridging. Fig. 2, p. 17.

Kenya. Kwale District: Kilibasi Hill, Nov. 1989, *Luke & Robertson* 2052! & fl. in hort. Luke, Aug.
1997, *Luke* 4705!
Tanzania. Arusha District: Loitang, Ngurdoto Crater, Oct. 1965, *Greenway & Kanuri* 12109!;
Lushoto District: W Usambara Mts, Lushoto-Mkuzi road near Magamba fork, Apr. 1953,
Drummond & Hemsley 2131!; Morogoro District: Uluguru North Forest Reserve, ridge above
Maunga Valley W of Bondwa, July 1972, *Mabberley* 1211!

© The Board of Trustees of the Royal Botanic Gardens, Kew, 2006

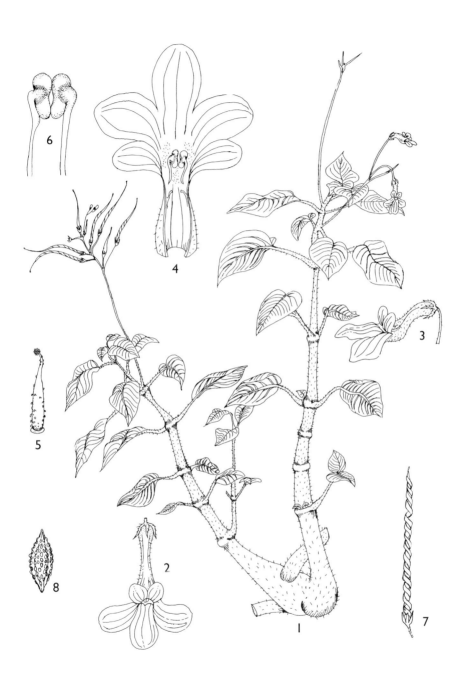

FIG. 2. *STREPTOCARPUS PALLIDIFLORUS* — **1**, habit × ²/₃; **2**, corolla, view from above × 2; **3**, corolla, lateral view × 2; **4**, dissected corolla showing position of stamens and staminodes × 4; **5**, pistil × 3.5; **6**, detail of stamens × 10; **7**, dehisced capsule × 2; **8**, seed × 33. 1, 2, 3, 4, 6 from cult. Kew, s.n.; 5 from *Drummond & Hemsley* 2131; 7 from *Schlieben* 12156, 8 from *Harris* 197. Drawn by Riziki Kateya.

© The Board of Trustees of the Royal Botanic Gardens, Kew, 2006

Distr. **K** 7; **T** 2, 3, 6, 7; not known elsewhere
Hab. Rock crevices or on banks in forest or unshaded areas, rarely epiphytic; 750–2100(–2750) m
Uses. Cultivated as an ornamental
Conservation notes. This species is recorded as locally common at several collecting localities
    within its range and is adaptable to a range of habitats including forest edges and open rock
    outcrops, thus is unlikely to be threatened by agricultural encroachment. It is therefore
    assessed as of Least Concern (LC)

Syn. *S. caulescens* Vatke var. *pallescens* Engl. in E.J. 19: 154 (1894), Hilliard & B.L. Burtt,
        Streptocarpus: 333, t. 15 (1971). Type as for *S. pallidiflorus* C.B.Clarke
      S. caulescens Vatke fa. *pallescens* (Engl.) Engl., P.O.A.: 363 (1895)
      [*S. caulescens sensu* Hook.f. in Bot. Mag. 111: t. 6814 (1885) & *sensu* B.L. Burtt in K.B.: 82
        (1939) pro parte, *non* Vatke]

Note. Flower colour in this species is somewhat variable. Most widespread are those with a pale
    mauve corolla striped with purple lines on the limb (particularly the upper lip). However, in
    the W Usambaras and at Kilibasi in Kenya, where the flowers are at the higher end of the size
    range, they are deeper purple and often lack the striping, at least on the lobes of the lower
    lip; these thus resemble in colour the flowers of *S. caulescens* Vatke, from which they are
    distinguished by the conspicuously ventricose tube.
      The two collections from Kilibasi in Kenya, both derived from the same population, are
    interesting in that the pedicels are rather densely pubescent, the peduncles and calyces also
    being more pubescent than typical for this species, although this is most notable in the
    cultivated specimen (*Luke* 4705!). If further collections from this area prove this to be a
    consistent character, it may warrant subspecific status.
      One of the original syntypes of this taxon, *Hannington* s.n.!, is unusual in that the ovaries
    are sparsely hairy which C.B. Clarke used to aid separation of this species from *S. caulescens*
    Vatke. However, this character is not repeated in subsequent collections, all having glabrous
    ovaries excepting the scattered subsessile glands. In all other respects, *Hannington* s.n. closely
    resembles specimens included under *S. pallidiflorus*, and as variable indumentum is well
    known in this genus, it is clear that they represent the same taxon.

9. **Streptocarpus caulescens** *Vatke* in Linnaea 43: 323 (1882); Baker & C.B. Clarke
in F.T.A. 4(2): 507 (1906); B.L. Burtt in K.B.: 81 (1939); Hilliard & B.L. Burtt,
Streptocarpus: 332 (1971); Blundell, Wild Fl. E. Afr.: 380, t. 597 (1987); Iversen in
Symb. Bot. Upsal. 28: 238 (1988); Agnew, U.K.W.F. ed. 2: 264 (1994). Type: Kenya,
Teita District, Mt Ndara, *Hildebrandt* (B†, holo.)

Straggling caulescent herb, 20–50(–75) cm tall. Stems succulent, reddish, slightly
swollen at the base and the nodes, spreading-pubescent particularly when young.
Leaves opposite, pairs subequal; blade somewhat fleshy, ovate-elliptic, 3–8.5 cm long,
2–4.5 cm wide, apex acute, base asymmetrically rounded to shortly attenuate, margin
entire, surfaces densely pubescent particularly on the nerves beneath; lateral nerves
8–10(–12) pairs, conspicuous beneath, inconspicuous above; petiole 1.5–3(–4.5) cm
long, densely pilose. Inflorescences axillary, solitary, often less than 10(–20)-
flowered; peduncles 3.5–14 cm long, spreading-pubescent towards the base,
becoming largely glabrous towards the apex; pedicels 8–20 mm long, glabrous or
with sparse glandular hairs, with scattered subsessile glands; bracts linear, 1–4 mm
long, pubescent, often early caducous. Calyx lobes purplish, ovate-lanceolate, 2–3.5 mm
long, pubescent particularly towards the tips, with eglandular and occasional
glandular hairs and scattered subsessile glands. Corolla deep violet, sometimes paler
at the mouth, 12–20 mm long, pendulous; tube cylindric, 5–9 mm long, 2.5–3 mm
diameter, floor straight or rarely slightly ventricose, glandular-pilose outside,
papillose inside below the origin of the stamens and at the mouth; limb bilabiate,
oblique, upper lip of two erect, rounded lobes, 2–3.5 mm long, 3–4.5 mm wide, lower
lip protruding, 7–12 mm long, lateral lobes 3–4 mm long, 4.5–7 mm wide, median
lobe 4–5.5 mm long, 6–8 mm wide, all broadly rounded, palate strongly curved
upwards and ridged, almost closing the mouth. Stamens arising from the upper third
of the corolla tube; filaments straight, 2.5–3.5 mm long, glabrous; anther thecae
divergent, ± 0.8 mm wide; staminodes minute. Ovary cylindric, 4.5–5.5 mm long,

© The Board of Trustees of the Royal Botanic Gardens, Kew, 2006

glabrous except for scattered subsessile glands, rarely sparsely pubescent; style 2–3 mm long, glabrous; stigma capitate, ± 0.5 mm wide, papillose. Capsule (20–)35–45 mm long, 1.5–2.5 mm diameter, glabrous. Seeds 0.4–0.6 mm long, verruculose with slight longitudinal ridging.

KENYA. Machakos District: Mt Nzaui, June 1966, *Wood* 700!; Masai District: S Chyulu hills, July 1938, *Bally* 8078!; Teita District: Vuria Peak, Apr. 1960, *Verdcourt & Polhill* 2727!
TANZANIA. Lushoto District: W Usambara Mts, Shume, Oct. 1943, *Moreau* 901! & Vugiri Hill, July 1954, *Faulkner* 1449! & Gombelo Forest Reserve, Bumba area, Apr. 1987, *Borhidi et al.* 87051!
DISTR. **K** 4, 6, 7; **T** 3; not known elsewhere
HAB. Terrestrial, epiphytic or lithophytic in moist areas in forest or forest margins; 950–2100 m
USES. Cultivated as an ornamental
CONSERVATION NOTES. Although localised in distribution, this species is recorded by several collectors as being locally frequent to common in suitable habitat within Teita District. It appears from herbarium collections to be scarce elsewhere within its known range. The persistance of forest over 1000 m alt. in the Taita Hills therefore appears important to the future survival of this species, a factor in some doubt due to widespread exploitation of the forest there by man. This species is therefore provisionally assessed as Near Threatened (NT), but may qualify as Vulnerable under IUCN criterion A once more complete data on forest loss in Taita becomes available

SYN. *S. rivularis* Engl. in E.J. 18: 78 (1895); Baker & C.B. Clarke in F.T.A. 4(2): 508 (1906). Type: Lushoto District: Usambara Mts, Mlalo, *Holst* 342 (B†, holo.)

NOTE. Specimens from the W Usambara Mts have a shallowly ventricose floor to the corolla tube above the middle, thus somewhat resembling the sympatric *S. pallidiflorus.* However, the swelling of the corolla tube is less pronounced than in the latter species, and in other respects (the violet corolla with no streaking on the limb, the strongly upcurved and ridged palate almost closing the corolla mouth, the lobes of the lower corolla lip being broader than long) they resemble *S. caulescens* and are thus placed in this species here. *S. rivularis* is tentatively placed as a synonym of this taxon here on the basis that the protologue description and illustration broadly agree with existing collections of *S. caulescens* from this area. Further investigation of this taxon within the W Usambaras, particularly the study of live plants and flowers in alcohol, is desirable to fully elucidate the species delimitation, and particularly to determine whether hybridisation occurs with *S. pallidiflorus.*
  Gillett 16304!, from a plant cultivated in Kenya from seed collected in the Teita Hills, is unusual in having sparse eglandular and glandular hairs on some of the ovaries in addition to the usual scattered glands. In all other respects it resembles *S. caulescens* and is therefore included within this species. Variable indumentum is well known within this genus, though not previously within this species.

10. **Streptocarpus holstii** *Engl.* in E.J. 18: 77 (1893); Baker & C.B. Clarke in F.T.A. 4(2): 508 (1906); Hemsley in Bot. Mag. 83: t. 8150 (1907); B.L. Burtt in K.B.: 83 (1939); Hilliard & B.L. Burtt, Streptocarpus: 334, fig. 49 (1971); Iversen in Symb. Bot. Upsal. 28: 243 (1988). Type: Tanzania, Lushoto District, Usambara, Nderema, *Holst* 2233 (B†, holo.; K!, iso.)

Caulescent herb, 20–60 cm tall, much-branched, branches usually decumbent. Stems somewhat succulent, subangular, nodes sometimes slightly constricted in dry material, sparsely spreading-pubescent, more densely pubescent at the nodes. Leaves opposite, pairs subequal to unequal; blade broadly ovate to ovate-elliptic, base rounded or obtuse, ± asymmetric, margin entire, apex bluntly acute or subacuminate, ciliate, upper surface sparsely pilose with clearly multicellular hairs, occasionally glabrous, lower surface pilose on the nerves, sparsely pubescent elsewhere; lateral nerves 6–10 pairs; petiole 0.5–3.5 cm long, pubescent with mixed long and short hairs. Inflorescences axillary to the upper leaves, solitary, 2–6-flowered; peduncles 2.5–8(–10) cm long, glabrous except for scattered hairs towards the base; pedicels 8–25 mm long, glabrous or sparsely glandular-pilose, bent at the apex rendering the flowers pendulous; bracts linear, ± 1.5 mm long, pubescent, often

© The Board of Trustees of the Royal Botanic Gardens, Kew, 2006

early caducous. Calyx lobes purplish, oblong-lanceolate with a blunt green tip, 2–3 mm long, with or without scattered glandular and eglandular hairs. Corolla (18–)20–29 mm long, sparsely glandular-pubescent outside; tube blue-purple, cylindric, 9–12 mm long, slightly swollen at the base, slightly curved and laterally compressed, 3(–3.5) mm deep, papillose on the floor within; limb bilabiate, highly oblique; upper lip of two erect lobes, blue to blue-purple at the margins, white towards the base, rounded, 2–4 mm long, 3–5 mm wide; lower lip protruding, blue to blue-purple except for a white palate with deeper blue flecks towards the edges, 10–15(–20) mm long, lateral lobes rounded or obtuse, 4–6 mm long, 6–9 mm wide, median lobe rounded or with a somewhat truncate apex, narrowed at base, 4.5–6 mm long, 6–9 mm wide, palate strongly upcurved and two-ridged, the mouth closed except for the narrow groove between the palate ridges. Stamens arising from the upper third of the corolla tube, lying flat on tube floor; filaments white, converging, 3–3.5(–5) mm long, glabrous; anthers white, thecae somewhat divergent, 0.5–0.6 mm wide; staminodes minute. Ovary cylindric, 4.5–5.5 mm long, glabrous except for scattered subsessile glands; style white, 3–5 mm long, glabrous; stigma shallowly depressed centrally, ± 0.6 mm wide, papillose. Capsule 30–55 mm long, 1–2 mm diameter, glabrous. Seeds 0.45–0.6 mm long, verruculose.

TANZANIA. Lushoto District: E Usambara Mts, Amani, Sept. 1933, *B.D. Burtt* 4862! & Kwamkoro-Ngua, Dec. 1936, *Greenway* 4815! & Mt Bomole, May 1950, *Verdcourt* 183!
DISTR. **T** 3; restricted to the E Usambara Mts
HAB. Banks and rocks in rivers in moist forest, rarely epiphytic; 750–1250 m
USES. Cultivated as an ornamental
CONSERVATION NOTES. The highly localised distribution of this species and its requirement for primary forest renders it vulnerable to the widespread human encroachment in the east Usambaras. Only twelve wild collections have been seen despite extensive collection in these mountains, thus it appears scarce. It is therefore assessed as Vulnerable (VU B1ab(iii)+2ab(iii))

SYN. *S. caulescens* Vatke var. *ovatus* C.B.Clarke in Monogr. Phan. 5: 154 (1883). Type: Tanzania, Lushoto District, Usambara, Magila, *Kirk* (K!, holo.)
    S. ovatus (C.B.Clarke) C.B.Clarke in F.T.A. 4(2): 508 (1906)

11. **Streptocarpus glandulosissimus** *Engl.* in E.J. 18: 78 (1893); Baker & C.B. Clarke in F.T.A. 4(2): 509 (1906); B.L. Burtt in K.B.: 82 (1939); Hilliard & B.L. Burtt, Streptocarpus: 337 (1971); Troupin in Fl. Rwanda, Spermatophytes 3: 499, fig. 151 (1985); Blundell, Wild Fl. E. Afr.: 380, t. 598 (1987); Agnew, U.K.W.F. ed. 2: 264 (1994). Types: Tanzania, Lushoto District, Usambara Mts, Mlalo, *Holst* 99 (B†, syn.) & Usambara Mts, Mtai, Tewe-Bach, *Holst* 2472 (B†, K!, HBG, syn.)

Perennial caulescent herb, straggling or subscandent, to 150(–200) cm tall. Stems somewhat succulent, weak, sometimes constricted at the nodes in dried material, branching, pubescent when young, glabrescent except for the pilose nodes, stomata often visible as purple streaks. Leaves opposite, pairs somewhat unequal or subequal; blade variable, ovate to ovate-elliptic, 3.5–14 mm long, 2–6 cm wide, base shallowly cordate to rounded or obtuse, often highly asymmetric, becoming cordate on one side and obtuse on the other, margin entire, apex acuminate or more rarely acute, surfaces pilose; lateral nerves 8–15 pairs; petiole 1–4.5(–6) cm long, pilose. Inflorescences axillary to the upper leaves, solitary, 4–22-flowered; peduncles 5–16 cm long, eglandular-pilose at the base, with progressively more glandular hairs towards the apex; pedicels 6–20 mm long, glandular-pilose, often densely so, the hair-walls pigmented purple, with shorter eglandular hairs and subsessile glands; bracts linear-lanceolate, 2.5–5 mm long, pilose, often early caducous. Calyx lobes broadly lanceolate, 2.5–4.5 mm long, blunt-tipped, pilose with mainly eglandular hairs except towards the base where hairs mainly glandular, with scattered subsessile glands. Corolla blue-purple to purple, often with a paler or whitish mouth, 16–30 mm long, glandular-pilose outside particularly on the tube; tube cylindric, 7–13 mm long,

© The Board of Trustees of the Royal Botanic Gardens, Kew, 2006

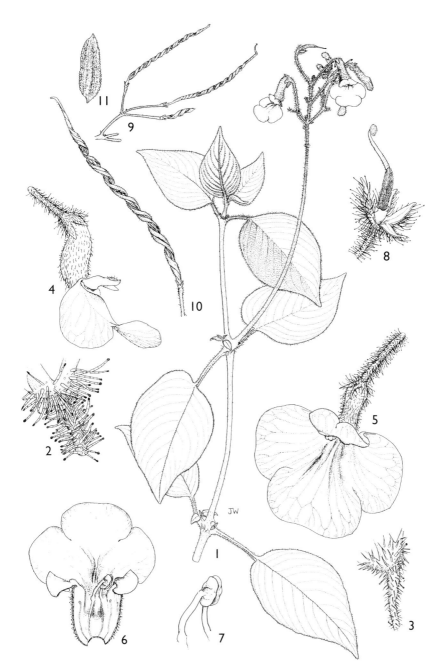

FIG. 3. *STREPTOCARPUS GLANDULOSISSIMUS* — **1**, habit × ²/₃; **2**, detail of pedicel and base of calyx showing glandular hairs × 8; **3**, detail of pedicel and base of calyx, variant with few glandular hairs × 8; **4**, corolla and calyx, lateral view × 2; **5**, corolla and calyx, front view × 2; **6**, dissected corolla showing position of stamens and staminodes × 2; **7**, detail of stamens × 5; **8**, detail of pistil and disk within calyx × 3; **9**, partial inflorescence in fruit × ²/₃; **10**, dehisced capsule × 1.5; **11**, seed × 40. 1 from *Pócs et al.* 88013/B; 2 from *Bigger* 2148; 3, 9, 10 from *Loveridge* 283; 4, 5, 6, 7, 8 from *Greenway & Kanuri* 11992; 11 from *Mhoro* in UMBCP 479. Drawn by Juliet Williamson.

© The Board of Trustees of the Royal Botanic Gardens, Kew, 2006

2.5–3(–4) mm deep, slightly swollen above the base, sometimes slightly ventricose on the floor of the upper half, papillose on the floor within; limb bilabiate, oblique; upper lip of two erect, rounded lobes, 3–5 mm long, 4–6 mm wide, united to midway; lower lip protruding, 12–20 mm long, lateral lobes rounded to asymmetric, 5–8(–10) mm long, 5.5–13.5(–15.5) mm wide, median lobe rounded or with a truncate to shallowly emarginate apex, 5–10.5 mm long, 7–17 mm wide, palate upcurved and two-ridged with a central groove, the mouth strongly vertically compressed, papillose. Stamens arising from the upper third of the corolla tube; filaments converging, 3–4 mm long, glabrous or with scattered short hairs; anther thecae divaricate, 0.8 mm wide; staminodes minute. Ovary cylindric, 4–6 mm long, densely appressed-pubescent with scattered subsessile glands; style 3–4 mm long, becoming glabrous towards the apex; stigma capitate, 0.6–1 mm wide, papillose. Capsule 35–60 mm long, 1.5–2 mm diameter, sparsely pubescent. Seeds 0.4–0.6 mm long, minutely verruculose with weak longitudinal ridges. Fig. 3, p. 21.

UGANDA. Ruwenzori, Mahoma, June 1953, *Osmaston* 3850! & Bwamba Pass, Nov. 1935, *A.S. Thomas* 1463! & Mubuku Valley, Dec. 1938, *Loveridge* 283!
KENYA. Northern Frontier District: Mt Kulal, Dec. 1958, *T. Adamson* K.20!; South Nyeri District: Castle Forest Station, Dec. 1966, *Kabuye* 41!; Kiambu District: Gatamayu Forest, Dec. 1959, *Napper* 1483!
TANZANIA. Moshi District: Kilimanjaro, S slope between the Umbwe and Weru Weru R., Aug. 1932, *Greenway* 3024!; Lushoto District: Usambara Mts, forest above Lushoto, July 1960, *Leach & Brunton* 10192!; Morogoro District: Uluguru Mts, Nyandiduma Forest Reserve, Mar. 1955, *Semsei* 2007!
DISTR. **U** 2; **K** 1, 4; **T** 2, 3, 6, 7; mountains of the central African lakes region in Congo-Kinshasa, Rwanda and Burundi
HAB. Terrestrial in damp shaded areas including riverbanks, pathsides and wet rocks in forest or forest margins, rarely epiphytic; (900–)1200–2250(–2600) m
USES. Cultivated as an ornamental
CONSERVATION NOTES. This is the most widespread species of *Streptocarpus* within our region and is often locally common to abundant within suitable habitat in the highlands of east and central Africa, though is notably absent from the Kenyan highlands west of the Rift Valley and from Mt Elgon. As *S. glandulosissimus* appears adaptable to a range of habitats and can survive in forest margins as well as pristine forest, it is not likely to experience threatening population declines at present and is thus assessed as of Least Concern (LC)

SYN. *S. volkensii* Engl. in E.J. 19: 153 (1894); Baker & C.B. Clarke in F.T.A. 4(2): 510 (1906). Type: Tanzania, Kilimanjaro, Marangu, *Volkens* 589 (B†, holo.; BM!, BR, K!, iso.)
   S. ruwenzoriensis Baker in F.T.A. 4(2): 510 (1906). Types: Uganda, Ruwenzori, *Scott-Elliott* 7936 (BM!, K!, syn.) & 7968 (BM!, syn.); Kenya, *Taylor* s.n. (BM!, syn.) & *Doggett* s.n. (K!, syn.)
   S. smithii C.B.Clarke in F.T.A. 4(2): 511 (1906). Type: Tanzania, Lushoto District, Usambara Mts, Umba Valley, *Smith* (K!, holo.)
   S. bequaertii De Wild. in Rev. Zool. Afr. 8, Suppl. Bot.: 38 (1920). Type: Congo-Kinshasa, Ruwenzori, Butahu [Batagu] valley, *Bequaert* 3554 (BR, holo.)
   S. mildbraedii Engl. in E.J. 57: 215 (1921). Type: Congo-Kinshasa, W Ruwenzori, Butahu [Butago] valley, *Mildbraed* 2497 (B†, holo.)
   S. tchenzemae Gilli in Ann. Naturhist. Mus. Wien 77: 54 (1973). Type: Tanzania, Morogoro District, Uluguru Mts, Chenzema, *Gilli* 549 (W!, holo.), **syn. nov.**

NOTE. A somewhat variable species, particularly with regard to pedicel indumentum. In the Ruwenzori and in some of the Kenyan collections, the pedicels are only sparsely glandular-pilose, with eglandular hairs dominating. Such specimens were previously considered a separate species, *S. ruwenzoriensis* Baker, but were correctly reduced to synonymy by Hilliard & B.L. Burtt (1971), as this and other diagnostic characters, notably the relative bluntness of the calyx lobes, are highly variable. The southernmost collections from Tanzania again display variable pedicel indumentum. Collections from the Udzungwa Mts (e.g. *Polhill & Lovett* 5155!, *Norbury* F27! & *Bridson* 604!) are particularly notable in displaying very sparse glandular hairs on the pedicels, but those from elsewhere in Iringa District are more typically densely glandular-pilose. The former specimens are somewhat difficult to separate from *S. buchananii* C.B.Clarke on inflorescence characters, but have small, ovate leaves with an acuminate apex, asymmetrically rounded base and rather dense indumentum which closely agree with other collections of *S. glandulosissimus* from Iringa.

© The Board of Trustees of the Royal Botanic Gardens, Kew, 2006

S. tchenzemae Gilli was separated from *S. glandulosissimus* on the basis of its taller stature with longer internodes and a longer corolla tube. Whilst the corolla tube length of 13 mm noted on the type specimen is at the longest end of the range for this species, it otherwise closely matches other specimens of *S. glandulosissimus* in indumentum, leaf characteristics and corolla form and is thus considered synonymous with that species.

12. **Streptocarpus buchananii** *C.B.Clarke* in F.T.A. 4(2): 510 (1906); B.L. Burtt in K.B.: 81 (1939); Hilliard & B.L. Burtt, Streptocarpus: 331 (1971); Cribb & Leedal, Mountain flow. south. Tanzania: 123, t. 30 (1982); Hilliard & B.L. Burtt in F.Z. 8(3): 59 (1988). Type: Malawi, Shire Highlands, *Buchanan* 410 (E, lecto., selected by Hilliard & B.L. Burtt (1988); K!, isolecto.)

Caulescent herb, 30–100 cm tall. Stems succulent, weak, sparsely pubescent mainly at the nodes, rarely branching. Leaves opposite, pairs subequal or unequal; blade ovate-elliptic, 7–14.5 cm long, 3–5.5 cm wide, base asymmetrically acute to shortly attenuate, margin entire, apex subacuminate, surfaces sparsely pubescent, often most dense on the nerves beneath; lateral nerves 10–16 pairs, ascending; petiole 1–6 cm long, pubescent. Inflorescences axillary to the upper leaves, solitary, 6–15-flowered; peduncles 8.5–18 cm long, puberulous; pedicels 15–30 mm long, puberulous with or without occasional longer glandular hairs; bracts linear, 1.5–2.5 mm long, pubescent, rarely with glandular hairs. Calyx lobes lanceolate, 2–3 mm long, pubescent with occasional glandular hairs, particularly towards the base, and with scattered subsessile glands. Corolla deep violet with a whitish mouth and inner tube, 17–23 mm long; tube cylindric, slightly curved, 9–12 mm long, 3 mm deep, glandular-pubescent outside, papillose at the mouth and below the origin of the stamens; limb bilabiate, oblique; upper lip of two rounded lobes, 2–4 mm long, 3–5 mm wide; lower lip 8–12.5 mm long, with two lateral lobes 4–5 mm long, 6.5–8 mm wide, median lobe 5–6 mm long, 6–7.5 mm wide, all rounded, palate curved upwards, the mouth strongly vertically compressed. Stamens arising from the upper half of the corolla tube; filaments straight, 3.3–4.4 mm long, glabrous; anther thecae divergent, 0.7 mm wide; staminodes minute. Ovary narrowly cylindric, 5–7 mm long, appressed pubescent, with scattered subsessile glands; style 3–4 mm long, pubescent towards the base; stigma capitate, 0.7 mm wide, papillose. Capsule slender, 50–80 mm long, 1.5(–2.5) mm diameter, sparsely pubescent to glabrescent. Seeds 0.5(–0.9) mm long, longitudinally ridged, verruculose.

TANZANIA. Morogoro District: Uluguru Mts, N Catchment Forest Reserve, above Tegetero Mission station, June 2000, *Jannerup & Mhoro* 91!; Rungwe District: Mwankinja, Mkinja river, Feb. 1954, *Semsei* 1625!; Mbeya District: Poroto Mts, Igale Pass between Igogwe and Igale, Feb. 1979, *Cribb et al.* 11258!
DISTR. **T** 6, 7; southern Tanzania and Malawi
HAB. Epiphytic or terrestrial in submontane to montane forest and forest edges, also by rivers; (1000–)1500–2200(–2600) m
USES. None recorded on herbarium specimens
CONSERVATION NOTES. This species has a somewhat localised distribution, though is widespread in Malawi where it is known from at least four broad locations, from mountains in both the north and south of the country. It is likely under-recorded due to its frequently epiphytic habit. In its northernmost range, the Uluguru Mts and northern Iringa District of Tanzania, it is perhaps under-collected due to its similarity to the much commoner *S. glandulosissimus* Engl. A provisional assessment of Data Deficient (DD) is therefore applied here, as more complete data on its abundance and range are required

SYN. *S. lilacinus* Engl. in E.J. 57: 214 (1921). Type: Tanzania, Rungwe, *Stolz* 558 (B†, holo.; BM!, K!, iso.)
   [*S. caulescens sensu* C.B.Clarke in Monogr. Phan. 5: 154 (1883) quoad *Buchanan* 410, *non* Vatke]

NOTE. This species is clearly closely allied to *S. glandulosissimus* Engl. and two collections from the Uluguru Mts (*Pócs* 6426/V! and *Harris* BJH 4548!) appear intermediate in character, suggesting that hybridisation may occur where their ranges overlap.

© The Board of Trustees of the Royal Botanic Gardens, Kew, 2006

13. **Streptocarpus inflatus** *B.L.Burtt* in Hilliard & B.L. Burtt, Streptocarpus: 389 (1971). Type: Tanzania, Iringa District, Mt Image, *Polhill & Paulo* 1619 (E, holo.; EA!, K!, iso.)

Caulescent herb to 30 cm tall. Lower stem creeping and rooting at the nodes, sometimes swollen, somewhat succulent, flowering stems decumbent, with scattered spreading hairs and subsessile glands. Leaves opposite, somewhat unequal to subequal; blade (ovate-)elliptic, 3–10 cm long, 1.5–5 cm wide, base asymmetrically obtuse to subcordate, margin entire, apex bluntly acute to subacuminate, surfaces sparsely pubescent mainly on the nerves beneath, ciliate; lateral nerves 6(–8) pairs, sometimes purple beneath; petiole 1–4 cm long, sparsely pubescent or glabrescent. Inflorescences axillary to the upper leaves, solitary, 2–10-flowered; peduncles 3–7 cm long, largely glabrous towards the base, becoming sparsely glandular- and eglandular-pubescent and with subsessile glands towards the apex; pedicels 3–6 mm long, with dense subsessile glands and less dense long glandular-hairs; bracts linear, 1–2 mm long, early caducous. Calyx lobes linear-lanceolate, bluntly tipped, 3.5–5.5 mm long, sparsely pubescent with mainly eglandular hairs, with scattered subsessile glands outside and within. Corolla 25–32 mm long; tube mauve, curved-cylindric, 14–17 mm long, laterally constricted, inflated on the roof above the base where 3 mm deep, narrowing to 1.5–2 mm, sparsely glandular-pubescent towards the mouth; limb bilabiate, oblique, lips blue darkening to purple towards centre, throat white; upper lip of two erect, rounded lobes, 5–6.5 mm long, 4–5 mm wide; lower lip protruding, 12–15.5 mm long, lateral lobes rounded or somewhat obtuse, 6–9 mm long, 8.5–10 mm wide, median lobe rounded, 7–10.5 mm long, 9.5–12 mm wide, palate curved upwards, mouth narrow vertically. Stamens arising from the upper third of the corolla tube; filaments 2–3 mm long, glabrous; anther thecae divergent, 0.8 mm wide; staminodes minute. Ovary cylindric, 4–7 mm long, glabrous except for scattered fine hairs and subsessile glands; style white, 4–6 mm long, glabrous; stigma white, capitate, 0.6–0.8 mm wide, papillose. Capsule 35–55 mm long, 1.5–2 mm diameter, glabrous. Seeds 0.6–0.8 mm long, minutely verruculose, slightly longitudinally ridged.

TANZANIA. Iringa District: Mt Image, Apr. 1986, *Congdon* 68! & Luhombero Massif, Aug. 1985, *Rodgers & Hall* 4375! & Udzungwa Mt National Park, May 2002, *Luke et al.* 8478!
DISTR. **T** 7; restricted to Mt Image and the Udzungwa Mts
HAB. Terrestrial, epiphytic or lithophytic in moist montane forest; 2100–2350 m
USES. None recorded on herbarium specimens
CONSERVATION NOTES. This species is clearly rare, being known from only six collections, three of which are from Mt Image. It has only recently been discovered at its second and third sites, both within the Udzungwa Mt National Park. No data on present abundance are available for its known locations or on the threats to these populations, although as it is found at such high altitudes, human impact upon its habitat may still be minimal. However, due to its highly restricted range, it is threatened by losses of subpopulations through natural stochastic events, such as forest fires, and is thus provisionally assessed as Vulnerable (VU D2)

14. **Streptocarpus euanthus** *Mansf.* in N.B.G.B. 12: 96 (1934); Hilliard & B.L. Burtt, Streptocarpus: 325 (1971). Type: Tanzania, Morogoro District, Uluguru Mts, Mkambaku, *Schlieben* 3589 (B†, holo.; BM!, BR!, MU, iso.)

Scrambling perennial caulescent herb to 2 m tall. Young stems shortly pubescent with mainly eglandular hairs, glabrescent. Leaves opposite, pairs somewhat unequal, falling from older stems leaving prominent scars; blade ovate, 2.5–4.8 cm long, 1.5–2.5 cm wide, base rounded to subcordate, slightly asymmetric, margin entire, apex subacuminate, densely pubescent above, principal nerves pubescent beneath; lateral nerves 9–11 pairs, conspicuous beneath; petiole 1–4 cm long, densely pubescent. Inflorescences axillary to the upper leaves, solitary, 2–7-flowered; peduncles 5–11.5 cm long, sparsely eglandular- and glandular-pubescent, most dense at the base, old inflorescences sometimes persisting at lower nodes, then peduncles

© The Board of Trustees of the Royal Botanic Gardens, Kew, 2006

glabrescent; pedicels 8–10 mm long, mainly eglandular-pubescent, with scattered glandular hairs; bracts not seen. Calyx lobes linear, 5–6(–10) mm long, sparsely eglandular-pubescent, with some glandular hairs and subsessile glands at the base. Corolla blue-lilac, 35–40 mm long; tube narrowly cylindric, 20–24 mm long, ± 2.5 mm deep, sparsely glandular-pubescent outside, mouth open; limb oblique; upper lip of two ovate lobes, 4–5 mm long, ± 5 mm wide; lower lip 14–17 mm long, lateral lobes ± 6 mm long, 8–9 mm wide, median lobe ± 8 mm long, 9 mm wide, all rounded. Stamens arising from the upper third of the corolla tube; filaments 2(–3) mm long, anther thecae rounded, ± 1 mm wide; staminodes not observed. Ovary narrowly cylindric, 13–16 mm long, glabrous, narrowing into the style, ± 10 mm long; stigma shallowly bilobed, ± 0.7 mm wide. Capsule 75(–90) mm long, ± 2 mm diameter, glabrous. Seeds not seen.

TANZANIA. Morogoro District: Uluguru Mts, NE side, Mkambaku, Feb. 1933, *Schlieben* 3589! (type)
DISTR. **T** 6; known only from the Uluguru Mts
HAB. Montane mist forest; ± 2280 m
USES. None recorded on herbarium specimens
CONSERVATION NOTES. This species has not been recollected in its only known site of the Uluguru Mts since its first discovery, despite several collecting expeditions into this area in the intervening seventy years. It is therefore clearly either extremely rare or extinct here, and qualifies as Critically Endangered (CR B1ab(iv)+2ab(iv)). A thorough search in its type location and surrounding suitable habitat in the Uluguru Mts is highly advisable to try to rediscover this species and, if successful, to establish an action plan for its survival

15. **Streptocarpus bambuseti** *B.L.Burtt* in K.B.: 80 (1939) & in Notes Roy. Bot. Gard. Edinb. 22: 574 (1958); Hilliard & B.L. Burtt, Streptocarpus: 326 (1971). Type: Tanzania, Morogoro District, Nguru Mts, Mafulumla, *Schlieben* 4094 (B†, holo., K!, photo.; BM!, BR, MU, iso.)

Erect caulescent herb or subshrub to 1 m tall. Stems woody at the base where sometimes shortly decumbent and branching; young stems densely pilose with glandular and eglandular hairs, reddish-tinged. Leaves opposite, pairs subequal to slightly unequal; blade elliptic, 6–11.5 cm long, 3–5 cm wide, often reddish-purple beneath, base asymmetric, shortly attenuate, margin entire, apex shortly acuminate, densely pilose above, less dense beneath except along the principal nerves; lateral nerves 7–12 pairs, obscured above; petiole 2–3 cm long, densely pilose. Inflorescences axillary to the upper leaves, solitary, lax, 2–16-flowered; peduncles 4.5–12 cm long, glandular-pilose, peduncles of old inflorescences often persisting at lower axils, then glabrescent and straw-coloured; pedicels 6–10 mm long, densely glandular- and eglandular-pilose; bracts lanceolate, ± 1.5 mm long, caducous. Calyx lobes lanceolate, 4.5–8 mm long, densely glandular- and eglandular-pubescent. Corolla blue-violet, paler at the mouth and within the tube, 40–60 mm long; tube narrowly cylindric, 25–37 mm long, ± 2 mm deep, sparsely pilose outside and around the open mouth; limb bilabiate, oblique; upper lip of two erect, rounded lobes, 7–8 mm long, 5.5–7 mm wide; lower lip 20–25 mm long, lateral lobes rounded-obovate, 10–11 mm long, 11–12 mm wide, median lobe obovate, 11–13.5 mm long, 10–13.5 mm wide, palate with two parallel dark violet streaks. Stamens arising near the apex of the corolla tube; filaments 2.5–4.5 mm long, pubescent; anthers purple, suborbicular, 2–2.5 mm wide, held at the corolla mouth, thecae indistinct. Ovary narrowly cylindric, 12–17 mm long, densely glandular- and eglandular-pubescent, narrowing into the style, 1.2–2.5 mm long, pubescent; stigma bilobed, 0.6–1.2 mm wide. Capsule 35–100 mm long, 2.5–3 mm diameter, glabrescent. Seeds ± 0.9 mm long, verruculose along longitudinal ridges.

TANZANIA. Morogoro District: Nguru Mts, saddle to NW of Mkobwe, near Turiani, Mar. 1953, *Drummond & Hemsley* 1907! & Ruhamba Peak, Apr. 1953, *Drummond & Hemsley* 1988! & Chazi Valley NW from Mhonda Mission, May 1989, *Pócs & Orban* 89163/M!
DISTR. **T** 6; known only from the Nguru Mts

© The Board of Trustees of the Royal Botanic Gardens, Kew, 2006

HAB. Terrestrial in moist montane and submontane forest; 1400–2000 m

USES. None recorded on herbarium specimens

CONSERVATION NOTES. This species is currently known from less than 10 specimens. It was not collected during the expedition in the Nguru Mts by Bidgood *et al.* in March 1988 despite this being within the species' flowering period. It may therefore be scarce, although it was found one year later by Pócs & Orban. Forest clearance is known to be occurring within this species' altitudinal range, thus habitat loss could threaten its future survival. It is therefore provisionally assessed as Endangered (EN B1ab(iii)+2ab(iii))

SYN. *S. glandulosissimus* Engl. var. *longiflorus* Mansf. in N.B.G.B. 12: 95 (1934). Type as for *S. bambuseti* B.L.Burtt

16. **Streptocarpus schliebenii** *Mansf.* in N.B.G.B. 12: 95 (1934); Hilliard & B.L. Burtt, Streptocarpus: 324 (1971). Type: Tanzania, Morogoro District, Nguru Mts, Mesumba, *Schlieben* 4186 (B†, holo.; BM!, BR!, MU, iso.)

Probably monocarpic, caulescent herb, 20–100 cm tall. Stems often decumbent, branched near the base where somewhat succulent, unbranched in the upper half, densely pilose with glandular and eglandular hairs. Leaves opposite, pairs ± unequal; blade oblong-elliptic to oblong-lanceolate, 8–14 cm long, 2.5–4 cm wide, base acute to obtuse, often asymmetric, margin crenate to crenate-serrate, apex shortly acuminate, surfaces ± densely pubescent, particularly above; lateral nerves 12–18 pairs; petiole 0.5–4.5 cm long, glandular-pilose. Inflorescences axillary to the upper leaves and ultimately terminal, solitary, 8–35-flowered; peduncles 2.5–7.5 cm long, pedicels 5–10 mm long, both densely pilose with glandular and eglandular hairs; bracts oblanceolate, 5.5–8 mm long, pubescent. Calyx lobes lanceolate, 1.5–3 mm long, glandular-and eglandular-pilose outside. Corolla white, sometimes with a pink spot at the base of the tube within, obliquely subcampanulate, 6–12 mm long, sparsely glandular-pubescent outside; tube 4–7 mm long, widening from a narrow base to ± 5 mm deep at the open mouth, ± densely pilose within, particularly on the roof; limb bilabiate; upper lip of two erect, rounded lobes, 2–2.5 mm long and wide, lower lip of three ovate-rounded lobes, ± 3 mm long, 3.5 mm wide. Stamens arising from near the base of the corolla tube; filaments slender, 2–3 mm long, slightly kinked, glabrous or sparsely pubescent; anther thecae 0.8 mm wide; staminodes minute. Ovary narrowly conical, 1.5–2 mm long, shortly glandular-pubescent or rarely largely glabrous; style 4–5 mm long, glabrous except at base; stigma rounded, 0.3–0.4 mm wide, papillose. Capsule 15–23 mm long, 1.5–2 mm diameter, tightly twisted, glandular-pubescent. Seeds 0.5 mm long, verruculose.

TANZANIA. Kilosa District: Ukaguru Mts, Mamiwa Forest Reserve, NNE slopes of Mnyera Peak, July 1972, *Mabberley et al.* 1279!; & E slope on Mamiwa Mt, May 1978, *Thulin & Mhoro* 2770!; Morogoro District: Nguru Mts, near Maskati Mission, Mabega Mt, June 1978, *Thulin & Mhoro* 3052!

DISTR. **T** 6; restricted to the Ukaguru and Nguru Mts

HAB. Terrestrial in deep shade in montane or submontane forest, occasionally lithophytic on shaded rocks; (1500–)1900–2200 m

USES. None recorded on herbarium specimens

CONSERVATION NOTES. Until the 1970s this species was known only from the type collection; exploration in the Ukaguru and Nguru Mts in that decade then revealed several further sites including four collections at two sites in the former, with Mamiwa Forest Reserve being a possible stronghold. Montane forest patches remain intact in both regions, but human encroachment is likely to increase in the future as much of the lower forest has already been cleared. This species is therefore provisionally assessed as Vulnerable (VU B2ab(iii))

17. **Streptocarpus parensis** *B.L. Burtt* in Notes Roy. Bot. Gard. Edinb. 22: 576 (1958); Hilliard & B.L. Burtt, Streptocarpus: 324 (1971). Type: Tanzania, Pare District, S Pare, Mt Shengena, *Peter* O II 53 (B!, holo.)

© The Board of Trustees of the Royal Botanic Gardens, Kew, 2006

Perennial caulescent herb to 20 cm tall or more. Stems decumbent, woody towards the base, densely glandular-pubescent in the upper section. Leaves opposite, pairs somewhat unequal; blade oblong-elliptic, 6–12 cm long, 2.5–3.5 cm wide, base attenutate, asymmetric, margin crenate-serrate, apex acute or subacuminate, upper surface pilose, lower surface pubescent mainly on the nerves, hairs eglandular and glandular; lateral nerves 17–18 pairs, conspicuous beneath; petiole 0.5–2 cm long, glandular-pubescent. Inflorescences axillary to the upper leaves, solitary, 10–15-flowered; peduncles 5–8 cm long, densely glandular-pubescent, peduncles of old inflorescences persisting at lower axils, then glabrescent; pedicels ± 5 mm long, densely glandular-pubescent; bracts linear, ± 3.5 mm long, glandular-pubescent. Calyx lobes oblong-lanceolate, 2–2.5 mm long, apex obtuse, outer surface pilose, inner surface with subsessile glands. Corolla white, obliquely subcampanulate, ± 10 mm long, sparsely glandular-pubescent outside; tube ± 7.5 mm long, widening to ± 4 mm deep at the open mouth, with long unicellular hairs on the roof within; limb with two upper lobes ± 1.5 mm long and wide, three lower lobes 2–2.5 mm long and wide, all rounded. Stamens arising almost at the base of the corolla tube; filaments slender, 2.5–3 mm long, slightly kinked, glabrous; anthers reniform, thecae 1 mm wide; staminodes to 0.75 mm long. Pistil not seen. Capsule 9–10 mm long, ± 2 mm diameter, tightly twisted, with sparse hairs or glabrate. Seeds 0.4 mm long, verruculose.

TANZANIA. Pare District: S Pare, Mt Shengena, Feb. 1915, *Peter* O II 53! (2 collections including type)

DISTR. **T** 3; restricted to the Pare Mts

HAB. On granite blocks in moist forest; 2150–2200 m

USES. None recorded on herbarium specimens

CONSERVATION NOTES. This species has not been re-recorded in the Pare Mts since its first discovery there. This is however possibly due to the under-exploration of this area rather than a reflection of the species' scarcity. Although significant areas of forest remain in these mountains, illegal logging is said to be occurring (K. Vollesen, *pers. comm.*), thus its habitat may be threatened. It is thus provisionally assessed as Vulnerable (VU D2) but may qualify as Critically Endangered if direct evidence of habitat threat is found

NOTE. This species is closely related to *S. schliebenii*, from which it is isolated geographically. It differs most notably in its habit, being a perennial species as evidenced by the presence of persistent old inflorescences at the lower leaf axes and by the woody base to the stem. *S. schliebenii* appears monocarpic, the inflorescences being axillary but eventually suppressing the terminal axillary bud and thus appearing terminal. *S. schliebenii* also lacks the presence of old inflorescences at the lower nodes.

18. Streptocarpus sp. A

Caulescent herb. Stems creeping towards the base, decumbent, pilose and with occasional shorter glandular hairs. Leaves alternate; blade oblong-ovate, 8–10.5 cm long, 3.5–4.5 cm wide, base attenutate, asymmetric, margin serrate, apex subacuminate, surfaces pubescent, particularly towards the margin above and on the nerves beneath, hairs eglandular; lateral nerves 10–12 pairs; petiole 3.5–6.5 cm long, pilose with occasional glandular hairs. Inflorescences axillary to the upper leaves, solitary, few-flowered; peduncles ± 9.5 cm long, pubescent, the hairs of variable length, mainly eglandular; pedicels 4–11 mm long, pubescent with longer glandular and shorter eglandular hairs; bracts oblanceolate, 1.7–3 mm long, pilose. Calyx lobes lanceolate, 1.5–2 mm long, outer surface pilose with occasional glandular hairs towards the base. Corolla white, obliquely subcampanulate, 9–12 mm long, glandular-pubescent outside; tube ± 7 mm long, widening to ± 4 mm deep at the open mouth, with long unicellular hairs on the roof within; limb with two upper lobes ± 2 mm long and wide, three lower lobes 1.8–2.7 mm long, 1.7–2.5 mm wide, all rounded. Stamens arising almost from the base of the corolla tube; filaments ± 1.7 mm long, slightly kinked, glabrous; anthers rounded, thecae ± 0.8 mm wide; staminodes minute. Ovary cylindric, ± 1.8 mm long, glandular-pubescent; style ± 2.5 mm long, glabrous except at base; stigma shallowly bilobed, ± 0.7 mm wide, papillose. Capsule short, spirally twisted. Seeds not seen.

© The Board of Trustees of the Royal Botanic Gardens, Kew, 2006

Tanzania. Lushoto District: Shagayu Forest, *Procter* 179! & Mazumbai Forest Reserve, Dec.
    1980, *Hall* MAZ 102! & ibid., Aug. 1982, *Hall* MAZ 226!
Distr. **T** 3; restricted to the Usambara Mts
Hab. Damp shaded rocks; 1850–1980 m
Uses. None recorded on herbarium specimens
Conservation Notes. With only two known sites, this species is likely to prove vulnerable once
    its taxonomy is fully delimited.

Syn. [*S. parensis sensu* Iversen, SAREC Usambara rain for. proj. report: 36 (1988) quoad *Procter*
    179, non B.L.Burtt]

Note. The rather depauperate *Procter* specimen, from which the above description is derived,
    was noted as possibly being referable to *S. parensis* by B.L. Burtt (in Hilliard & B.L. Burtt,
    Streptocarpus: 324 (1971)). However, several differences to that taxon are notable: the
    Procter specimen is a less robust plant lacking woody lower stems, the leaves are apparently
    alternate, proportionally broader and more ovate, have a serrate, not crenate-serrate margin
    and have fewer lateral nerve pairs, the petioles are longer, and the plant is generally less
    hairy, in particular having fewer glandular hairs on the vegetative parts. The two gatherings
    from Mazumbai Forest Reserve, both housed at DSM, have been seen only briefly by the
    author and are currently unavailable for more detailed analysis. They are again rather scant,
    and are smaller plants than the *Procter* collection but otherwise appear to agree with it.
    Formal description of this new species awaits further analysis of that material.

19. **Streptocarpus montanus** *Oliv.* in Trans. Linn. Soc. Bot. 3: 344 (1887); Baker &
C.B. Clarke in F.T.A. 4(2): 507 (1906) as *S. montana*; Hilliard & B.L. Burtt, Streptocarpus:
216, fig. 37 (1971); Iversen in Symb. Bot. Upsal. 28: 238 (1988); Agnew, U.K.W.F. ed. 2:
264 (1994). Type: Tanzania, Kilimanjaro, *H.H. Johnston* 157 (K!, holo.; BM!, iso.)

Acaulescent perennial herb. Rhizome creeping, stout, to 4(–6) mm thick, softly
pubescent, with prominent leaf scars. Leaves tending to cluster towards rhizome apex,
usually rosulate though occasionally solitary; blade narrowly oblong-elliptic or oblong-
lanceolate, size variable, 7–36 cm long, 1.5–17.5 cm wide, base ± asymmetric, cuneate
to acute, margin coarsely dentate, apex subacuminate or acute, often withered, densely
pubescent particularly on the upper surface and nerves beneath; lateral nerves 19–35
pairs, parallel, spreading; petiole 0.5–2(–4) cm long. Inflorescences apparently axillary,
solitary to several arising from the leaf axis, somewhat lax and spreading, 8–many-
flowered; peduncles 4–29 cm long; pedicels 7–17(–28) mm long, glandular- and
eglandular-pubescent; bracts linear, 2–4.5 mm long, pubescent. Calyx lobes lanceolate,
1.5–3.5 mm long, glandular- and eglandular-pubescent. Corolla pale- or less commonly
medium-violet, rarely whitish, with violet stripes on the floor of the tube, 11.5–19 mm
long, scattered glandular-pubescent outside; tube obliquely cylindric with a declinate
floor, 7–11 mm long, 5–7.5 mm deep, mouth open; limb bilabiate; upper lip of two
suberect, rounded lobes, 2.5–3 mm long, ± 2.5 mm wide; lower lip of three spreading,
oblong-rounded lobes, 4–4.5 mm long, 2–3.5 mm wide. Stamens arising from near the
base of the corolla tube; filaments U-curved, slender, 4–6 mm long, glabrous; anther
thecae rounded, 0.7–0.8 mm wide; staminodes minute. Ovary 3–4.5 mm long,
glandular-pubescent; style 3–4 mm long; stigma bilobed, 0.35–0.4 mm wide, papillose.
Capsule 15–25 mm long, 1–1.5 mm diameter, glandular-pubescent. Seeds 0.5–0.7 mm
long, verruculose, slightly ridged longitudinally. Fig. 4, p. 29.

Kenya. South Nyeri District: Mt Kenya, S side, Kamweti track above Castle Forest Station, Jan.
    1971, *Faden et al.* 71/102!; Masai District: Namanga, Ol Donyo Orok, June 1974, *Archer* 751!;
    Teita District: Vuria Peak, Apr. 1960, *Verdcourt & Polhill* 2728!
Tanzania. Moshi District: Kilimanjaro, S slope between Umbwe and Weru Weru R., Aug. 1932,
    Greenway 3031!; Pare District: N Pare Mts, Minja Forest Reserve, above Vuchama Ngofi village,
    Apr. 1990, *Pócs* 90064/J!; Morogoro District: S Nguru Mts, Ruhamba Peak, Apr. 1953,
    Drummond & Hemsley 1981!
Distr. **K** 4, 6, 7; **T** 2, 3, 6; restricted to the mountains of southern Kenya and eastern Tanzania
Hab. Epiphytic, lithophytic or terrestrial in shaded areas in wet montane or submontane forest;
    (1200–)1500–2400(–2800) m

© The Board of Trustees of the Royal Botanic Gardens, Kew, 2006

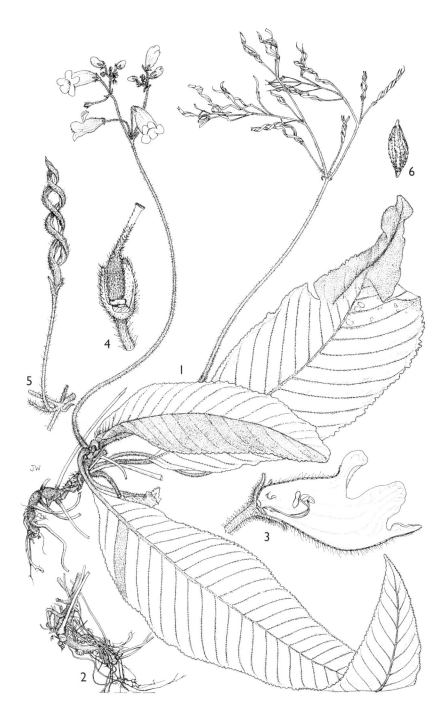

FIG. 4. *STREPTOCARPUS MONTANUS* — **1**, habit × ²/₃; **2**, detail of rhizome × ²/₃; **3**, longitudinal cross section of flower, showing stamens and one staminode, pistil removed × 4; **4**, pistil within partially dissected calyx × 5; **5**, dehisced capsule × 2; **6**, seed × 24. 1 from *Verdcourt & Polhill* 2728; 2, 5, 6 from *Faden* 71/102; 3 from *Greenway* 9722, 4 from *Drummond & Hemsley* 1857. Drawn by Juliet Williamson.

© The Board of Trustees of the Royal Botanic Gardens, Kew, 2006

USES. None recorded on herbarium specimens

CONSERVATION NOTES. Although of restricted range, this species is locally common in several mountain ranges, for example forming large colonies on mountains in the Taita Hills. It is adaptable to a range of niches within its habitat and is tolerant of a range of climatic conditions, as evidenced by its large altitudinal range. Those populations at high altitudes are likely to experience little human disturbance at present. It is therefore assessed as of Least Concern (LC)

NOTE. A specimen cultivated at R.B.G. Edinburgh, labelled "grown from seed collected by *Wallace* in Tanganyika Territory" (K!) is close to *S. montanus* but differs in having a larger paniculate inflorescence with up to 80 flowers which are only 8.5–10 mm long and have a proportionally narrower tube than in *S. montanus*. These characters place it superficially similar to *S. umtaliensis* B.L.Burtt from the Zimbabwe-Mozambique borderlands but that species has the stamens inserted higher up the corolla tube and largely lacks glandular hairs on the inflorescence. The filaments of the stamens are often almost straight in this specimen, though some display the U-bend typical of *S. montanus*. The material is incomplete, as the petiolode and rhizome (if present) have not been pressed. It also lacks locality notes and it cannot even be confirmed that the material originated from Tanzania, although Wallace is known to have collected primarily in the Uluguru Mts (K. Vollesen, *pers. comm.*). Further material is needed of this taxon which may prove to be either a variant of *S. montanus* or a closely allied new species.

20. **Streptcarpus sp. B**

Acaulescent herb, with or without a short rhizome. Unifoliate to bifoliate, the second leaf much reduced; principal leaf pendant, blade ovate, 7.3–8.5 cm long, 4.7–5.5 cm wide, base asymmetric, rounded, margin coarsely bidentate, apex acute to obtuse, surfaces densely pubescent, particularly above and on the nerves beneath; lateral nerves 12 pairs, parallel, shortly ascending; petiole to 0.5 cm long, pubescent. Infructescence apparently axillary, solitary, fruits paired on the specimens observed, though scarring on the peduncle suggests more flowers may have originally developed; peduncle 6.5–8.5 cm long, with scattered hairs; fruiting pedicels 10–17 mm long, glabrescent; bract remnants pubescent. Calyx, corolla and stamens not seen. Capsule 25–33 mm long, glabrous. Seeds 0.45–0.5 mm long, verruculose, slightly ridged longitudinally.

TANZANIA. Iringa District: Udzungwa Mt National Park, Sanje–Mwanihana route, Nov. 1997, *P.A. & W.R.Q. Luke* 5065!
DISTR. **T** 7; not known elsewhere
HAB. Rock faces; 1130 m
USES. None recorded on herbarium specimens
CONSERVATION NOTES. If this proves to be a good species, it clearly has a highly limited distribution and will likely qualify as threatened within IUCN criteria

NOTE. Known only from the collection cited, this taxon may be a dwarf form of *S. montanus*, though it differs in the proportionally broader leaf blade with characteristic bidentate margin, in having a much-reduced or absent rhizome and in having short peduncles with few fruits. *S. montanus* is currently unknown from the Udzungwa Mts. Flowering material is required to resolve this potentially new taxon.

21. **Streptocarous burttianus** *Pócs* in Fragm. Flor. Geobot. Ann. 35(1–2): 40 (1991). Type: Tanzania, Morogoro District, Nguru Mts, Dikurura Valley, W of Mhonda Mission, *Pócs, LaFarge-England & Magill* 90057/A (BP, holo.; K!, SUA, iso.)

Perennial acaulescent herb. Rhizome short or inconspicuous, pubescent. Flowering plants usually bifoliate with one large leaf blade, oblong, oblong-ovate or obovate, 20–50(–100) cm long, 12–25 cm wide, base truncate to attenuate, margin dentate, apex obtuse, often withered, surfaces appressed-pubescent, most dense along the nerves beneath, upper surface shiny in living material; lateral nerves 30–40(–50) pairs, parallel, spreading; second leaf much reduced or absent, to 6 cm long; petiolode to 1.5 cm long, densely appressed-pubescent. Inflorescences

© The Board of Trustees of the Royal Botanic Gardens, Kew, 2006

apparently axillary, solitary to several, subcorymbiform, 10–40-flowered; peduncles 7–13 cm long, (sub-)appressed eglandular-pubescent, peduncles of old inflorescences sometimes persisting, then glabrescent; pedicels 2.5–9 mm long, glandular- and eglandular-pubescent; bracts lanceolate, 1.5–3 mm long, pubescent particularly beneath. Calyx lobes narrowly lanceolate, 2–3 mm long, densely eglandular-pubescent. Corolla white with yellow markings in the throat, subcampanulate, 7–11 mm long, shortly pubescent outside and at the mouth with mainly eglandular hairs; tube with a declinate floor and straight roof, 5–7 mm long, (3–)4–6 mm deep just below the open, rounded mouth; limb slightly bilabiate; upper two lobes erect, rounded, 2–3 mm long, 2.5–3.5 mm wide, lower lip with a short palate, lateral lobes rounded, 3–3.5 mm long, 3–4 mm wide, median lobe subtriangular, 3–4.5 mm long, 3–5 mm wide. Stamens arising from near the base of the corolla tube; filaments strongly inflexed near the apex, 2–3 mm long, glabrous; anther thecae rounded, 0.7–0.9 mm wide; staminodes minute. Ovary narrowly conical, 2–2.5 mm long, densely appressed eglandular-pubescent, narrowing into the style, 2.5–3.5 mm long, becoming glabrous towards the apex; stigma capitate, 0.35–0.4 mm wide, papillose. Capsule 10–12 mm long, only slightly twisted, glabrescent. Seeds not seen.

TANZANIA. Morogoro District: Nguru Mts, Chazi valley NW from Mhonda Mission, May 1989, *Pócs & Orban* 89163/L! & Dikurura Valley, W of Mhonda Mission, Mar. 1990, *Pócs et al.* 90057/A! (type) & Dikurura Valley, Apr. 1989, *Farkas et al.* 89254!
DISTR. **T** 6; known only from the Nguru Mts
HAB. Shaded cliff faces in montane moist forest; 1100–2000 m
USES. None recorded on herbarium specimens
CONSERVATION NOTES. This species is highly localised in distribution, though it is noted in the protologue that the leaves form "large, shiny masses from shady cliffs in the montane rainforest belt of (the) Nguru Mountains" suggesting that it may be locally numerous. If this is the case, it is unusual that the species was first discovered in the late 1980s, though this may be due more to the paucity of previous collections from this mountain range than to the species' rarity. Being currently known from only three sites, it is vulnerable to localised loss of habitat, either through forest clearance by humans, or through stochastic events, the most likely being rockfalls. It is therefore provisionally assessed here as Vulnerable (VU D2)

NOTE. This species appears quite isolated in evolutionary terms in east Africa, sharing closest affinity with a central African species, *S. burundianus* Hilliard & B.L.Burtt from north of Lake Tanganyika in Burundi. *S. burttianus* however lacks the lateral flowering shoots with paired phyllomorphs characteristic of this species and the related *S. masisiensis* De Wild.

22. **Streptocarpus bullatus** *Mansf.* in N.B.G.B. 12: 96 (1934); Hilliard & B.L. Burtt, Streptocarpus: 216, t. 11, fig. 37 (1971). Type: Tanzania, Morogoro District, Uluguru Mts, Mkambaku mist forest, *Schlieben* 3586 (B†, holo., K!, illustr.; BM!, BR!, LISC, Z, iso.)

Perennial caulescent herb. Rhizome shortly creeping, 5–8 mm thick, pubescent, producing a series of erect flowering shoots; petiolode stem-like, 2.5–17 cm long, densely eglandular-pubescent sometimes with longer gland-tipped hairs, with a large leaf at the apex opposite which develops a much-reduced leaf whose blade may be carried upwards by extension of its petiolode, this pattern sometimes repeated so that up to five reduced leaves develop. Blade of lowest leaf oblong-elliptic, 5–21 cm long, 2–7.5 cm wide, often bullate, lower surface sometimes tinged purple, base cordate to obtuse, margin crenate to crenate-dentate, apex obtuse, both surfaces appressed-pubescent, on the upper surface the hairs often directed away from both the midrib and lateral nerves, on the lower surface most dense along these; lateral nerves conspicuous, 18–33 pairs, scalariform. Inflorescences solitary or up to three developing from below the base of the lamina, spreading, 2–20-flowered; peduncles 3–7 cm long; pedicels 2–9 mm long, wiry, both spreading-pubescent, with or without

© The Board of Trustees of the Royal Botanic Gardens, Kew, 2006

gland-tipped hairs; bracts oblong-lanceolate, 1.5–3 mm long, pubescent. Calyx lobes broadly lanceolate, 1.5–2.8 mm long, pubescent, with or without gland-tipped hairs. Corolla white, sometimes tinged or marked pink or purple, obliquely subcampanulate, 6–12.5 mm long, with scattered hairs outside and with numerous unicellular hairs on the roof of the tube and the inside of the upper two lobes; tube 4–5.5 mm long, 4.5–5 mm diameter, floor downcurved to the open mouth; limb subregular, upper two lobes suberect, rounded, 2–3.5 mm long, 3–4.5 mm wide, lower three lobes rounded, 2.5–4 mm long, 3–5 mm wide. Stamens arising from near the base of the corolla tube; filaments 1.5–2.5 mm long, almost straight or inflexed, glabrous; anther thecae ± 0.75 mm wide; staminodes minute. Ovary narrowly conical, 1–1.5 mm long, pubescent and/or with stalked glands; style 3–4.5 mm long, pubescent particularly towards the base; stigma rounded or shallowly bilobed, 0.2–0.4 mm wide, papillose. Capsule 5–14 mm long, 1.5–2 mm diameter, sparsely pubescent. Seeds ± 0.6 mm long, verruculose.

TANZANIA. Morogoro District: Uluguru Mts, near Chenzema on way to Lukwangula Plateau, Mar. 1955, *Semsei* 1961! & Mkumbaku Mt, June 1978, *Thulin & Mhoro* 3189! & above Tegetero Mission Station, June 2000, *Jannerup & Mhoro* 93!
DISTR. **T** 6; restricted to the Uluguru Mts
HAB. Moist, mossy rocks in forest; (1200–)1700–2250 m
USES. None recorded on herbarium specimens
CONSERVATION NOTES. Although highly restricted in range, this species can be locally "very common" in suitable habitat (*Semsei* 1961) in the Uluguru Mts. With eleven specimens seen, it is clearly more numerous than the following three species from the same mountain range. However, populations at the lower end if its altitudinal range may be threatened by loss of habitat, thus this species is provisionally assessed as Vulnerable (VU B1ab(iii)+2ab(iii))

SYN. *S. minutiflorus* Mansf. in N.B.G.B. 12: 97 (1934). Type: Tanzania, Morogoro District, Uluguru Mts, Mkambaku mist forest, *Schlieben* 3585 (B†, holo., K!, illus.; BM!, BR, LISC, Z, iso.)

NOTE. Due to the complex habit of upwards extension of the petiolode during development of the phyllomorphs, this species can appear truly caulescent with alternately-arranged sessile leaves, though with the initial blade disproportionately larger than the upper ones.

23. **Streptocarpus albus** (*E.A.Bruce*) *I.Darbysh.* **comb. nov.** Type: Tanzania, Morogoro District, Uluguru Mts, Lupanga Peak, Tanana, *B.D. Burtt* 3470 (K!, holo.)

Caulescent herb. Stems short, creeping towards base with nodal rooting, pubescent when young, glabrescent. Leaves alternate, subrosulate; blade sometimes purplish beneath, ± broadly oblong-elliptic, 3–10.5 cm long, 2–5 cm wide, base truncate, obtuse or cuneate, margin crenate-serrate, apex obtuse to rounded, upper surface with short to long appressed hairs, lower surface pubescent mainly on the nerves; lateral nerves 8–14 pairs, ascending, impressed above particularly when young; petiole 0.5–2 cm, long-pubescent. Inflorescences axillary, 1–2 per axil, 4–12(–16)-flowered; peduncles 4–8 cm long, pilose, hairs ascending to spreading, sometimes with scattered glandular hairs; pedicels 4–9(–13) mm long, ascending-pubescent to pilose, with or without glandular hairs; bracts ovate to obovate, 1.5–4.5 mm long, with long hairs mainly on the margin. Calyx lobes (oblong-) lanceolate, (1.5–)2–3(–3.5) mm long, pilose, with scattered glandular hairs. Corolla white or rarely mauve-tinged, somewhat drooping, obliquely subcampanulate, 10–13.5 mm long, sparsely pubescent outside with glandular and/or eglandular hairs, glabrous within; tube 5–7 mm long, floor downcurved to the open, papillose mouth; limb with two suberect upper lobes, 2.5–3 mm long, 2–2.5 mm wide, lateral lobes 4 mm long, 3.5 mm wide, median lobe 4–5 mm long, 3.5–4.5 mm wide, all rounded and with unicellular marginal hairs. Stamens arising from near the base of the corolla tube; filaments 2–3 mm long, glabrous; anther thecae rounded, 0.8–1.1 mm wide; staminodes to 0.75 mm. Ovary ovoid, 1.4–2 mm long, finely pubescent, the hairs ± dense, with few to many glandular hairs; style 3.5–4.5 mm long, pubescent at the

© The Board of Trustees of the Royal Botanic Gardens, Kew, 2006

base, becoming glabrous towards the apex; stigma rounded, 0.3–0.5 mm wide. Capsule not or slightly twisted, cylindric, 7–12 mm long, 1.5–2 mm diameter, pubescent to glabrescent, style persistent. Seeds ± 0.3 mm long, verruculose.

a. subsp. **albus**

Leaf blade 5.5–10.5 cm long, 2.5–5 cm wide, base obtuse to cuneate, hairs on upper surface rather short, 3–4-celled; lateral nerves 10–14 pairs, barely impressed above in mature leaves. Peduncles and pedicels eglandular, hairs mainly ascending and held against the stalk. Corolla always white. Ovary densely eglandular-pubescent with or without scattered glandular hairs.

TANZANIA. Morogoro District: Uluguru Mts, Lupanga Peak, Dec. 1933, *Michelmore* 860! & Nguru ya Ndege, Mar. 1981, *Beesan* 384 & Lupanga Peak, June 1983, *Polhill & Lovett* 4931!
DISTR. **T** 6; known only from the Uluguru and Nguru Mts
HAB. Moist montane forest in deep shade; 1900–2150 m
USES. None recorded on herbarium specimens
CONSERVATION NOTES. This subspecies is known with certainty from only two sites, where it is restricted to the high cloud forest. It was recorded in 1933 as being "locally common" at Lupanga Peak (*Michelmore* 860), but there are no more recent data on its population status. A sterile collection, quite likely of this taxon, was made at Kilangala (*Pócs & Mwanjabe* 6464/Y!) but fertile material is required for confirmation. With such a limited distribution and with a habitat requirement for dense primary forest, this subspecies is considered threatened by habitat loss and is thus provisionally assessed as Endangered (EN+2ab(iii))

SYN. *Saintpaulia alba* E.A.Bruce in K.B.: 475 (1933)
   Linnaeopsis alba (E.A.Bruce) B.L.Burtt in Gard. Chron., ser. 3, 72: 23 (1933); Weigend in Flora 195: 49, fig. 2 (2000)

b. subsp. **edwardsii** (*Weigend*) I.Darbysh. **comb. nov.** Type: Tanzania, Morogoro District, Uluguru Mts, summit of Palata, *Pócs et al.* 87182/C (K!, holo.)

Leaf blade 3–5.5 cm long, 2–3.7 cm wide, base truncate to obtuse, hairs on upper surface long, 5–6-celled; lateral nerves 8–10 pairs, strongly impressed above even in mature leaves together with tertiary venation, giving leaves a bullate surface. Peduncles and pedicels with both ascending and spreading hairs, eglandular and glandular, the latter most dense on the pedicels. Corolla white to pale mauve. Ovary thinly glandular-pubescent, sometimes glabrous.

TANZANIA. Morogoro District: Uluguru Mts, Bondwa Peak, Nov. 1962, *Moors* s.n.! & NW Uluguru Mts, on the Palata track, July 1972, *Mabberley* 1254! & summit of Palata, July 1987, *Pócs et al.* 87182/C! (type)
DISTR. **T** 6; known only from the Uluguru Mts
HAB. Moist montane forest; (1550–)1900–2000 m
USES. None recorded on herbarium specimens
CONSERVATION NOTES. This subspecies is again highly localised in range, currently known from only two sites. Little information on its abundance is available, although only one colony was encountered during botanical work in the northwest Ulugurus in 1972 (*Mabberley* 1254). The paucity of collections suggests that it is scarce, and as it requires dense primary forest it is threatened by localised habitat loss, thus is provisionally assessed as Endangered (EN B1ab(iii)+2ab(iii))

SYN. *Linnaeopsis alba* (E.A.Bruce) B.L.Burtt subsp. *edwardsii* Weigend in Flora 195: 49, fig. 2 (2000)

24. **Streptocarpus heckmannianus** (*Engl.*) I.Darbysh. **comb. nov.** Type: Tanzania, Morogoro District, Uluguru Mts, Lukwangule Plateau, *Goetze* 251 (B†, holo., K! photo.; K!, iso.)

Caulescent herb. Stems thin, creeping, producing fine adventitious roots, pubescent with long and/or short hairs. Leaves alternate; blade broadly ovate to suborbicular, 1–5.5 cm long, 1–4 cm wide, base cordate, truncate or obtuse, margin crenate, apex rounded or obtuse, surfaces pubescent, hairs on upper surface variable, long or short, most dense towards the margins, hairs on the lower surface most dense on the principal

© The Board of Trustees of the Royal Botanic Gardens, Kew, 2006

nerves; lateral nerves 4–12 pairs, ascending; petiole 0.7–3.5 cm long, pubescent. Inflorescences axillary, solitary, 2–8-flowered; peduncles 5–8.5 cm long, pilose with mainly or entirely eglandular hairs; pedicels 5–9(–13) mm long, pilose with mixed glandular and eglandular hairs, ± dense; bracts elliptic to oblanceolate, 1–4 mm long, pubescent. Calyx lobes (oblong-)lanceolate, 2–3.5(–4) mm long, eglandular-pubescent, sometimes with glandular hairs at the base. Corolla white, rarely tinged pink, somewhat drooping, obliquely subcampanulate, 10–12 mm long, sparsely pubescent outside, glabrous within; tube 4.5–7 mm long, floor downcurved to the open, papillose mouth; limb with two suberect upper lobes, 2.5–3.2 mm long, 2.5 mm wide, lateral lobes 4–5 mm long, 3.5–4.5 mm wide, median lobe 4.5–5 mm long, 3.5–5 mm wide, all rounded and with unicellular marginal hairs. Stamens arising from the base of the corolla tube; filaments 2–2.5 mm long, glabrous; anther thecae rounded, ± 1 mm wide; staminodes minute. Ovary ovoid, 2–2.2 mm long, glabrous to densely pubescent, hairs eglandular or glandular; style 4–5 mm long, largely glabrous away from base; stigma rounded, 0.3–0.5 mm wide. Capsule cylindric, 5–8 mm long, ± 1.5 mm diameter, not twisted, pubescent to glabrescent. Seeds 0.35 mm long, verruculose.

a. subsp. **heckmannianus**

Stem hairs mainly short. Leaves suborbicular, 1–2 cm long, 1–2.3 cm wide, base cordate; lateral nerves 4–5 pairs. Pedicels with mainly glandular hairs. Corolla always white. Ovary densely glandular-pubescent.

TANZANIA. Morogoro District: Uluguru Mts, Lukwangule Plateau, Nov. 1898, *Goetze* 251! (type) & Aug. 1933, *Schlieben* 4237! & Mgeta R., on path from Chenzema to Nyingwa, Dec. 2005, *Haston* 110!
DISTR. **T** 6; known only from the Uluguru Mts
HAB. Moist montane forest, riverside rocks; 1800–2300 m
USES. None recorded on herbarium specimens
CONSERVATION NOTES. This subspecies is currently known from only a narrow altitudinal range in the Uluguru Mts. Only three collections are known to the author despite significant botanical exploration in this mountain range. Its forest habitat is highly threatened in the Uluguru Mts where deforestation has been widespread and is ongoing. It is thus provisionally assessed as Endangered (EN B1ab(iii)+2ab(iii))

SYN. *Linnaeopsis heckmanniana* Engl. in E.J. 28: 484 (1900); Baker & C.B. Clarke in F.T.A. 4(2): 512 (1906); Weigend in Flora 195: 49, fig. 2 (2000)

b. subsp. **gracilis** (*E.A.Bruce*) *I.Darbysh.* **comb. nov.** Type: Tanzania, Morogoro District, Uluguru Mts, Matombo Road, Tanana, *E.M. Bruce* 802 (K!, lecto., selected by Weigend in Flora 195(1): 49 (2000); BM!, isolecto.)

Stem hairs mainly long. Leaves broadly ovate or more rarely suborbicular, (2–)3.5–5.5 cm long, (2.2–)2.5–4 cm wide, base truncate to obtuse; lateral nerves 6–12 pairs. Pedicels with mixed eglandular and glandular hairs, sometimes rather sparse, the latter usually with small gland tips. Corolla white or pink-tinged. Ovary glabrous or densely eglandular-pubescent.

TANZANIA. Morogoro District: Uluguru Mts, Matombo Road, Tanana, Feb. 1935, *E.M. Bruce* 802! (type) & Mkambaku Mt, June 1978, *Thulin & Mhoro* 3190! & Palu, Dec. 1993, *Kisena* 980!
DISTR. **T** 6; known only from the Uluguru Mts
HAB. Moist forest, including amongst rocks in forest streams; (1350–)1700–1950 m
USES. None recorded on herbarium specimens
CONSERVATION NOTES. This subspecies is now known from three sites within the Uluguru Mts, and from a wider altitudinal range than the typical subspecies, but still appears scarce and highly restricted in range. It again requires primary forest and is thus vulnerable to local forest disturbance. It is therefore provisionally assessed as Endangered (EN B1ab(iii)+2ab(iii))

SYN. *Linnaeopsis gracilis* E.A.Bruce in K.B. 9: 486 (1936)
    Linnaeopsis heckmanniana Engl. subsp. *gracilis* (E.A.Bruce) Weigend in Flora 195: 49, fig. 2 (2000)

© The Board of Trustees of the Royal Botanic Gardens, Kew, 2006

NOTE. *Kisena* 980! is somewhat intermediate in leaf size and shape between subsp. *gracilis* and subsp. *heckmannianus*, but is placed within the former on the basis of the glabrous ovary and the sparse glandular hairs on the pedicels, which have smaller, less conspicuous gland tips than in subsp. *heckmannianus*. Further collections may however reveal that the two subspecies currently recognised in fact represent the extremes of a single variable entity.

25. **Streptocarous subscandens** (*B.L.Burtt*) *I.Darbysh.* **comb. nov.** Type: Tanzania, Morogoro District, Uluguru Mts, Lupanga Peak, *Eggeling* 6273 (EA, holo.; K!, iso.)

Subscandent or scrambling herb. Stems robust towards the base, densely appressed-pubescent. Leaves alternate; blade ovate, 3–6 cm long, 1.7–5 cm wide, base obtuse or shortly attenuate, margin serrate, apex shortly acuminate or acute, surfaces shortly appressed-pubescent, most dense on the margins and nerves beneath; lateral nerves 5–8 pairs, ascending, conspicuous beneath; petiole (1–)1.5–3 cm long, densely appressed-pubescent. Inflorescences axillary, solitary, 2–12-flowered; peduncles 3–5(–7) cm long; pedicels 6–9(–12) mm long, both appressed eglandular-pubescent, the latter also with ± dense shorter, somewhat spreading, glandular hairs; bracts linear, 2.5–5 mm long, appressed-pubescent. Calyx lobes narrowly lanceolate, 3–4 mm long, appressed eglandular-pubescent, sometimes with glandular hairs towards the base. Corolla white, somewhat drooping, obliquely subcampanulate, 10–14 mm long, glabrous to sparsely pubescent outside particularly on the lobe margins, hairs short, usually glandular; tube with floor downcurved, ± 5.5 mm long, 7 mm deep at the open, rounded mouth, glabrous within except for short papillae around the mouth; limb with two erect, upper lobes, 3.5–4.5 mm long, 3–4 mm wide, lateral lobes 5–6 mm long, 4–5 mm wide, median lobe 4.5–7 mm long, 3.5–6.5 mm wide, all rounded. Stamens arising from the base of the corolla tube; filaments 1.5(–2) mm long, glabrous; anther thecae rounded, ± 1 mm wide, staminodes not observed. Ovary ovoid, ± 2 mm long, densely short glandular-pubescent; style 3–4 mm long, glandular-pubescent at base, becoming glabrous; stigma rounded, ± 0.2 mm wide, barely wider than the style. Immature capsule only seen, purple, subcylindric, slightly twisted. Seeds not seen.

TANZANIA. Morogoro District: Uluguru Mts, NW side, Lupanga Peak, Nov. 1932, *Schlieben* 2936! & Aug. 1951, *Eggeling* 6273! (type) & N Uluguru Forest Reserve, Ng'ubabule, Dec. 1993, *Kisena* 897!
DISTR. **T** 6; known only from the Uluguru Mts
HAB. Moist forest; 1500–2000 m
USES. None recorded on herbarium specimens
CONSERVATION NOTES. This taxon is known only from the three specimens cited, despite several collecting expeditions in this mountain range. It is therefore likely scarce and highly localised and so vulnerable to local losses of forest through human activity. It is provisionally assessed as Endangered (EN B1ab(iii)+2ab(iii))

SYN. *Linnaeopsis subscandens* B.L.Burtt in Notes Roy. Bot. Gard. Edinb. 22: 581 (1975); Weigend in Flora 195: 49, fig. 2 (2000)

NOTE. *Kisena* 897! is unusual in having leaves at the lowest end of the size range and having shorter, broader calyx lobes than the other two specimens. However, in the material seen the calyx is somewhat poorly preserved, thus the latter character may be an artefact of collection and drying; in all other respects it agrees closely with the type collection.

26. **Streptocarpus exsertus** *Hilliard & B.L.Burtt*, Streptocarpus: 389 (1971); Blundell, Wild Fl. E. Afr.: 380, t. 117 (1987). Type: Kenya, Turkana District, near Kesai, Chemorongit Mts, *J. Wilson* 1336 (EA!, holo.; E!, iso.)

Unifoliate herb. Leaf and inflorescence borne on a straight, stout petiolode, 0.5–3(–4.5) cm, densely covered with spreading, straw-coloured hairs. Leaf blade broadly ovate to oblong, (3–)7–9.5 cm long, (1.5–)4–11 cm wide, base shallowly

© The Board of Trustees of the Royal Botanic Gardens, Kew, 2006

cordate, margin obscurely crenulate to denticulate, apex withered in all material seen, surfaces ± densely pubescent, particularly on the nerves beneath; lateral nerves 7–9 pairs, ascending. Inflorescences several, arising from the apex of the petiolode, to 10 cm tall, several-flowered; peduncles slender, short, 0.5–2.7 cm long before branching to the secondary peduncles; pedicels slender, (20–)35–50 mm long, extending to 60 mm in fruit, pilose with narrow glandular and eglandular hairs, and with sessile or short-stalked glands; bracts absent in the material seen. Calyx lobes linear, 3–5 mm long, pilose with slender glandular and eglandular hairs. Corolla white, 9.5–14 mm long, glabrous or with scattered glandular hairs on the outside; tube cylindric, straight, 4–8 mm long, 2–3 mm diameter, glabrous within, mouth open, papillose on the upper rim; limb bilabiate; upper lip of two oblong lobes (1.5–)3–4 mm long, 1–2 mm wide; lower lip 4–8 mm long, the three lobes oblong-spathulate, (2–)4–6 mm long, 1–2 mm wide, palate densely papillose towards the edges. Stamens arising from the upper third of the corolla tube; filaments 2–3 mm long, slender, glabrous or with few sessile glands; anthers exserted from corolla tube or held at the mouth, thecae rounded, ± 0.5 mm wide; staminodes minute, arising from the lower third of the corolla tube. Ovary cylindric, 2.5–3.5(–6) mm long, ± glandular-pubescent, abruptly narrowed at the apex; style white, 4–5 mm long, sparsely pubescent, the hairs eglandular towards the apex; stigma white, bilobed, 0.5 mm wide, somewhat papillose. Capsule 10–17 mm long, 1–1.5 mm diameter, sparsely pubescent to glabrescent, not or only slightly twisted. Seeds 0.25–0.3 mm long, reticulate.

KENYA. Northern Frontier District: Lolokwi Mt, Apr. 1979, *Gilbert* 5358! & ibid., June 1995, *Bytebier* 564!; Turkana District: near Kesai, Chemorongit Mts, July 1963, *J. Wilson* 1336! (type)
DISTR. **K** 1, 2; restricted to the isolated mountains of northern Kenya
HAB. Moist rock faces and under rock outcrops, where it may experience periodic drought; 1450–1750 m
USES. None recorded on herbarium specimens
CONSERVATION NOTES. This species was recorded as very rare in the Chemorongit Mts in the type collection; its current status there is unknown. On Lolokwi Mt, Gilbert recorded it as locally common in one location in 1979, but Bytebier found "only one group of about 50 plants" on his visit to this site in 1995. Other isolated mountains occur in the intervening area between the two known locations, thus it may occur at other suitable sites within its range. At present, the isolation and apparantly small population sizes at the known locations suggest that it may be threatened by stochastic events such as extended drought and it is therefore assessed as Vulnerable (VU D2)

27. **Streptocarpus rhodesianus** *S.Moore* in J.B. 49: 188 (1911); B.L. Burtt in K.B.: 78 (1939); Hilliard & B.L. Burtt, Streptocarpus: 254 (1971); Hilliard & B.L. Burtt in F.Z. 8(3): 59 (1988). Type: Zambia, Katenina Hills, *Kassner* 2162 (BM!, holo.; K!, iso.)

Perennial acaulescent herb. Leaves (1–)several, rosulate; blade often purplish beneath, sessile, oblong (-lanceolate), 2–15(–30) cm long, 1–6(–9) cm wide, base subcordate to cuneate, margin entire, apex usually withered, surfaces densely pilose, the hairs long and white, often obscuring the surface; lateral nerves 8–20(–30) pairs, parallel and spreading. Inflorescences 1–several arising from the base of the leaves, (1–)2–6(–12)-flowered; peduncles 2.5–9 cm long, often thickened towards the base, with scattered long white eglandular hairs and with sparse to numerous glandular hairs towards the apex, the peduncles of previous seasons' inflorescences partially persisting; pedicels 2.5–12(–23) mm long, glandular-pubescent and with occasional long eglandular hairs; bracts oblong, 0.5–1.5 mm long, inconspicuous, caducous. Calyx lobes lanceolate, 1.3–3 mm long, apex blunt, outer surface with predominantly glandular hairs at the base and scattered long eglandular hairs towards the apex. Corolla with a mauve tube outside, the lips and tube within white, the latter with purple markings on the floor, 8.5–19 mm long, glandular-pubescent outside; tube subcylindric, 5.5–11 mm long, 2–4.5 mm diameter, broadening somewhat to the open mouth, the floor with long, blunt-tipped, sometimes clavate hairs, these

© The Board of Trustees of the Royal Botanic Gardens, Kew, 2006

extending onto the palate of the lower lip, the roof with finer pilose hairs within; limb spreading, bilabiate; upper lip of two rounded lobes, 1–4 mm long and wide; lower lip 3–8.5 mm long, lateral lobes 2–5 mm long, 1–5 mm wide, median lobe 2–5.5 mm long, 1–6 mm wide, all rounded to oblong-rounded. Stamens arising from the lower third of the corolla tube; filaments 2.5–3.5 mm long, centrally kinked, glabrous; anthers reniform, thecae rounded, 0.6–0.9 mm wide; staminodes minute, arising below the stamens. Ovary narrowly conical, 1.5–3 mm long, sparsely glandular- or eglandular-pubescent or rarely glabrate; style 2.5–4.5 mm long, sparsely pubescent towards the base; stigma shallowly bilobed, 0.4–0.6 mm wide, papillose. Capsule 5–12 mm long, ± 1.5 mm diameter, sparsely pubescent or glabrate, style persistent. Seeds 0.25–0.5 mm long, reticulate.

SYN. *S. paucispiralis* Engl. in E.J. 15: 217 (1921). Type as for *S. rhodesianus* S.Moore
    S. rhodesianus S.Moore var. *perlanatus* P.A.Duvign. in B.S.B.B. 96: 180 (1963). Type: Congo-
    Kinshasa, Katanga, Dikuluwe, *Duvigneaud* 5126 (BRLU, holo.)

a. subsp. **grandiflorus** *I.Darbysh.* **subsp. nov.** subspeciei typicae affinis sed floribus maioribus 15–19 mm (nec 8.5–14.5 mm) longis, lobis proportione longioribus e comparatione labio inferiore, et seminibus maioribus 0.5 mm (nec 0.25–0.35 mm) longis differt. Type: Tanzania, Songea District, Mbinga, *Congdon* 161 (K!, holo. & iso.)

Calyx lobes 2.3–3 mm long. Corolla 15–19 mm long; lower lip 6–8.5 mm long, lateral lobes 3.7–5 mm wide, median lobe 4–6 mm wide, all rounded and with an irregular margin; hairs on floor of corolla tube blunt but not clearly clavate. Capsule 8–12 mm long. Seeds 0.5 mm long.

TANZANIA. Songea District: Mbinga, Kiteza Forest Reserve, Mar. 1987, *Congdon* 161! (type)
DISTR. **T** 8; not known elsewhere
HAB. Exposed rock faces with thin soils; 1650 m
USES. None recorded on herbarium specimens
CONSERVATION NOTES. *S. rhodesianus* is a widespread species, the typical subspecies occuring from eastern Angola and southwestern Congo-Kinshasa through much of Zambia. As it grows around bolders and rock faces, its habitat is unlikely to be threatened by man. It is therefore assessed as of Least Concern (LC). Subsp. *grandiflorus* is represented by a single collection and may be restricted to the Kiteza Forest Reserve of southern Tanzania. With such a small distribution, it is vulnerable to stochastic events such as rock falls which may impact greatly upon the size of existing populations. It is therefore assessed as Vulnerable (VU D2)

NOTE. Subsp. *grandiflorus* superficially looks distinct from *S. rhodesianus* due largely to the strikingly large flowers with broad lobes to the lower lip. However, the two taxa are otherwise very close and share the striking indumentum of the inner corolla tube. The single collection of subsp. *grandiflorus* additionally displays small leaves (to only 6 cm long) and peduncles bearing only 1–2 flowers. Both characters are sometimes recorded in subsp. *rhodesianus*, but this latter taxon usually has larger leaves and more than two flowers on at least some of the peduncles.

28. **Streptocarpus solenanthus** *Mansf.* in N.B.G.B. 12: 96 (1934); Hilliard & B.L. Burtt, Streptocarpus: 185, t. 6, fig. 27 (1971); Cribb & Leedal, Mountain flow. south. Tanzania: 124, t. 30 (1982); Hilliard & B.L. Burtt in F.Z. 8(3): 47 (1988); Burrows & Willis, Pl. Nyika Plateau: 180, t. 3 (2005). Type: Tanzania, Ulanga District, Sali, *Schlieben* 1878a (B†, holo.)

Monocarpic acaulescent herb. Unifoliate, leaf often pendant; blade ovate to oblong, 6–35 cm long, 3.5–21 cm wide, base cordate, rarely attenuate, margin shallowly crenate-serrate, apex usually withered, spreading-pubescent, particularly on the upper surface and nerves beneath; lateral nerves 11–22 pairs, parallel, spreading, impressed above, raised beneath; petiolode less than 1 cm long, pubescent. Inflorescences solitary or up to 4(–7), arising from the petiolode, erect, 5–40-flowered, panicles subcorymbiform or rarely more elongate; peduncles 3–15(–22) cm long, eglandular-pubescent, the hairs in the lower part longer and

© The Board of Trustees of the Royal Botanic Gardens, Kew, 2006

deflexed; pedicels 7–22 mm long, shortly pubescent; bracts ovate-lanceolate to oblanceolate, 4–6(–9) mm long, pubescent. Calyx lobes oblong-lanceolate, 1.5–3.5 mm long, shortly pubescent on both surfaces. Corolla white to pale lavender-blue, the latter sometimes restricted to the inner limb, 22–35 mm long, shortly eglandular-pubescent outside; tube cylindric, 17–30 mm long, (3–)4–5.5 mm deep, scarcely widened to the open mouth, slightly downcurved towards the apex, glabrous within; limb only slightly oblique, lobes small in relation to tube, spreading; upper lip of two rounded lobes, 2–4.5 mm long, (2–)3–5.5 mm wide; lower lip 5–10 mm long, lobes rounded, 3–6 mm long, (2.5–)4–6 mm wide. Stamens arising from the upper third of the corolla tube; filaments 3–4 mm long, thickened centrally, glabrous; anther thecae subtriangular to rounded, 0.75–1 mm wide, pale purple-tinged; staminodes minute. Ovary cylindric, 11–16 mm long, densely shortly eglandular-pubescent, the hairs somewhat spreading; narrowing into the style, white, 4–8 mm long, pubescent; stigma shallowly bilobed, 0.9–1.2 mm wide, papillose. Capsule slender, 40–80 mm long, 1–2 mm diameter, sparsely pubescent. Seeds 0.4–0.65 mm long, reticulate.

Tanzania. Morogoro District: Chigurufumi Forest Reserve, Mar. 1955, *Semsei* 2025!; Iringa District: Udzungwa Mts, Kigogo Forest, Mar. 1988, *Bidgood et al.*, 853!; Songea District, Lupembe Hill, Mar. 1956, *Milne-Redhead & Taylor* 8960!

Distr. **T** 6–8; southern Tanzania to northern Malawi and eastern Zambia with an isolated population in eastern Zimbabwe

Hab. Epiphytic in forest or lithophytic on shaded or exposed rock faces; (1150–)1350–2100 m

Uses. None recorded on herbarium specimens

Conservation notes. This species is somewhat local in distribution. However, it has significant populations in several highland areas of southern Tanzania, northern Malawi and eastern Zimbabwe, appearing rare only in eastern Zambia. It is an adaptable species, being at home in both closed forest as an epiphyte, and in exposed, rocky environments. The latter are largely unthreatened by anthropogenic factors, thus this species is assessed as of Least Concern (LC)

Note. Corolla colour in this species displays significant geographic variation. Within our region, plants from the northern and eastern parts of the range, for example in Morogoro and Ulanga Districts, have exclusively white corollas, whereas those from further south, such as in Songea District tend to be pale lavender-blue, at least on the limb, the tube often remaining whitish.

29. **Streptocarpus sp. C**

Acaulescent unifoliate herb. Leaf subsessile; blade oblong-ovate, 7.5–10.5 long, 6.5–9 cm wide, base subcordate, margin obscurely crenulate, apex withered in material seen, surfaces pubescent, the hairs longest on the upper surface and nerves beneath; lateral nerves 8–11 pairs below the withered portion, parallel, spreading. Inflorescences 1–3, arising from the base of the leaf blade; peduncles 13–18 mm long; pedicels 5–10(–18) mm long, both densely eglandular-pilose; bracts linear, 3.5–5 mm long, pubescent, hairs longer above. Calyx lobes lanceolate, 2.5–3 mm long, densely pilose outside, short-pubescent within. Corolla white, ± 24 mm long, eglandular-pilose outside; tube cylindric, ± 17.5 mm long, 3–3.5 mm deep at the centre, expanding slightly towards the open mouth where 7 mm deep; limb with rounded upper lobes ± 4 mm long, 5 mm wide; lower lip ± 8 mm long, lobes rounded-obovate, 5–5.5 mm long and wide. Stamens arising from midway along the corolla tube; filaments ± 2 mm long, with short-stalked glands towards the apex; anther thecae rounded, ± 1 mm wide; staminodes not seen. Ovary narrowly cylindric, ± 7 mm long, densely spreading eglandular-pubescent, narrowing into the style, 3–3.5 mm long, short-pubescent; stigma shallowly bilobed, ± 1 mm wide, papillose. Capsule and seeds not seen.

Tanzania. Iringa District: Mt Selegu, N side, ± 20 km W of Mahange village, Feb. 1987, *Lovett & Congdon* 1471!

Distr. **T** 7; not known elsewhere

© The Board of Trustees of the Royal Botanic Gardens, Kew, 2006

Hab. Moss-covered rocks in woodland; 1400 m
Uses. None recorded on herbarium specimens
Conservation notes. If this proves to be a good species, it clearly has a highly limited
    distribution and will likely qualify as threatened under IUCN criteria.

Note. The single specimen cited differs from *S. solenanthus* in the stamens being positioned
    midway along the tube rather than in the upper third, the shorter ovary and style and the
    longer indumentum of the inflorescence. In addition, whereas the corolla tube is at the lower
    end of the range for *S. solenanthus* in this specimen, the lobes are at the larger end of the
    range, thus the limb appears significantly larger in relation to the tube than in *S. solenanthus*.
    However, the material currently available is scant, thus further collections are required.

30. **Streptocarpus mbeyensis** *I.Darbysh.* **sp. nov.** *S. solenanthus* similis sed corolla
maiore atque purpurea (nec alba neque lilacina), labio inferiore magis distincte
protruso, stylo longiore et stigmate latiore differt. *S. michelmorei* similis sed tubo corollae
proportione longiore e comparatione labio inferiore, stylo longiore et pilis glandulosis
e inflorescentia carentibus differt. Type: Tanzania, Mbeya District, Pungulumo
[Pungaluma] Hills, *Lovett, Sidwell & Kayombo* 4016 (MO!, holo.; BR!, K!, iso.)

Monocarpic acaulescent herb. Unifoliate; blade broadly ovate, 17–40 cm long,
12–30 cm wide, base shallowly cordate, margin crenate-dentate to crenate, apex obtuse,
usually withered, surfaces pubescent, hairs most dense and longest on the principal
nerves beneath; lateral nerves 14–16 pairs, parallel, spreading; petiolode less than 1 cm
long, pubescent. Inflorescences 2–5, arising from the petiolode and base of the midrib,
13–25(–40)-flowered; peduncles of primary inflorescence 6.5–9.5(–13) cm long,
pubescent, the hairs somewhat deflexed towards the base; pedicels 10–22 mm long,
spreading-pubescent, the hairs eglandular or rarely with a few scattered glandular hairs;
bracts linear to oblanceolate, 6–10 mm long, pubescent particularly beneath. Calyx
lobes oblong-lanceolate, 3–4(–5.5) mm long, eglandular-pubescent. Corolla purple, the
lobes darker, with or without a whitish patch under the throat outside, 33–44 mm long,
rather densely eglandular-pubescent outside, glabrous within; tube cylindric, 25–33 mm
long, 3.5–6 mm deep, slightly downcurved and somewhat expanded towards the open
mouth, where 8.5–10 mm deep; limb bilabiate; upper lip of two rounded lobes, 4–6 mm
long and wide; lower lip 7.5–14 mm long, lateral lobes 5–6(–8) mm long, 4.5–5.5
(–8.5) mm wide, median lobe 4.5–6(–9) mm long, 5–7(–8) mm wide, all rounded and
held forward. Stamens arising in the upper third of the corolla tube near the mouth;
filaments purple, 4–6.5 mm long, slightly thickened above the base, glabrous or with few
sessile glands towards the top; anther thecae purple and white, rounded, 1–1.5 mm
wide; staminodes arising from below the stamens, to 0.5 mm long. Ovary narrowly
cylindric, 11–19 mm long, densely pubescent, with short, spreading, eglandular hairs;
style 8–14 mm long, shortly pubescent, barely narrower than the ovary; stigma white,
shallowly bilobed, 1.7–2 mm wide, shortly papillose. Capsule 55–70 mm long, 1.5–2 mm
diameter, pubescent. Seeds ± 0.5 mm long, reticulate.

Tanzania. Mbeya District: Mbeya, May 1975, *Hepper & Field* 5502!; & below Ipota area, rock
    gorge leading down to Mbalizi river near Mshewe village, Jan. 1990, *Lovett et al.* 3886! &
    Pungolomo Hills, Jan. 1990, *Lovett et al.* 4016! (type)
Distr. **T** 7; not known elsewhere
Hab. Streamside forest or open woodland; 1200–1350 m
Uses. None recorded on herbarium specimens
Conservation notes. This species is known from only the three locations listed above, within
    a highly restricted range. It is found in gulleys or gorges of watercourses, within a variety of
    vegetation types from forest to rather dry woodland. One of the populations (*Hepper & Field*
    5502) was made from a residential area and is thus likely to be or have been lost through
    human encroachment. It is therefore considered Endangered (EN B1ab(iii) 2ab+(iii))

Note. This newly described species falls within *S.* aggregate *cooperi* C.B.Clarke *sensu* Hilliard &
    B.L.Burtt (in Streptocarpus: 175–186 (1971)), a complex of closely related species with long,
    largely cylindric corolla tubes which do not deepen strongly towards the mouth, and with
    (sub-)spreading hairs on the ovary. This group was previously represented in our region only

© The Board of Trustees of the Royal Botanic Gardens, Kew, 2006

by *S. solenanthus* from which *S. mbeyensis* is readily distinguished by the larger, darker corollas with a more clearly bilabitate limb, the longer style, the broader stigma and the somewhat coarser indumentum of the inflorescence. Of the other formally described taxa within this aggregate, *S. mbeyensis* is closest to *S. michelmorei* B.L.Burtt, a species restricted to the Zimbabwe-Mozambique borderlands. *S. michelmorei* differs in having a somewhat shorter corolla tube (20–27 mm long) with a proportionately longer limb (the tube being 2–2.4 times the length of the limb, this being 2.3–3 times in *S. mbeyensis*), in the stamens arising lower in the corolla tube and in having a shorter style (4.5–7 mm long, the length of 10 mm recorded by Hilliard & Burtt in F.Z. 8(3): 46 (1988), appearing to be erroneous). In addition, *S. michelmorei* often has numerous glandular hairs on the pedicels, corolla and ovary, though these are sometimes absent.

Hilliard & B.L. Burtt (in F.Z. 8(3): 46 (1988)) recognise two entities with close affinity to *S. michelmorei* which have never been formally described. The first (sp. 2A of F.Z.) is recorded from Chimaliro Hill on the Vipya Plateau. A specimen from the same location which closely matches the description of this taxon has been seen (*la Croix* 4353!); it is close to *S. mbeyensis* but differs in having a very narrow capsule (to 1.3 mm diameter), a puberulous indumentum to the inflorescence more akin to *S. solenanthus*, a less conspicuous stigma (1–1.3 mm wide) and a more slender corolla tube (3–4 mm diameter). The second (sp. 2B of F.Z.), recorded from northwestern and central Zambia (*Mutimushi* 3335! & *Robinson* 6630!, Kabompo Gorge; *Kornaś* 3166!, *Strid* 2896! & *Williamson* 1727! Kundalila Falls), appears to be the most closely allied taxon to *S. mbeyensis*. They share the very long corolla tube and pistil, purple corolla and long, slender capsule. The Zambian material however displays corollas with a proportionally smaller lower lip, the tube being 2.8–4.4 times its length and being more notably curved than in *S. mbeyensis*. In addition, the indumentum of the inflorescence is again more akin to *S. solenanthus*, being less coarse than in *S. mbeyensis*. Whether these differences are consistent remains unconfirmed, and the Zambian material may well prove conspecific with *S. mbeyensis*. A full revision of *S.* aggregate *cooperi* is desirable, ideally with the study of living wild plants; this may result in a broadened species concept. The extent to which hybridisation results in these highly localised taxa is also not fully understood and again requires further field knowledge. However, in the absence of such information, it is currently considered most appropriate to recognise the entities discussed above as a series of closely related but discrete species; as such *S. mbeyensis* is restricted to the Mbeya region of Tanzania.

31. **Streptocarpus goetzei** *Engl.* in E.J. 30: 406 (1901); B.L. Burtt in K.B.: 70 (1939); Hilliard & B.L. Burtt, Streptocarpus: 186, t. 7, fig. 27 (1971); Cribb & Leedal, Mountain flow. south. Tanzania: 124, t. 30 (1982); Hilliard & B.L. Burtt in F.Z. 8(3): 47 (1988); Burrows & Willis, Pl. Nyika Plateau: 180 (2005). Type: Tanzania, Njombe District, Livingstone Mountains, Yawuanda, *Goetze* 854 (B†, holo.; BR!, iso.)

Monocarpic acaulescent herb. Unifoliate, rarely plurifoliate in cultivation, leaf often pendant; blade ovate to oblong, size variable at flowering, 6–32 cm long, 2.5–30 cm wide, base shallowly cordate to truncate, margin crenate-dentate, sometimes obscurely so, apex usually withered, surfaces pubescent, the upper with long and sometimes short spreading hairs, the lower with long hairs along the nerves and shorter hairs inbetween; lateral nerves 12–20 pairs, parallel, spreading; petiolode usually less than 1 cm long, but to 2 cm, pubescent. Inflorescences solitary to 2–6(–10), erect, arising from the petiolode or base of the midrib, subcorymbiform, (3–)8–45-flowered; peduncles variable, (1.5–)7–26 cm long, eglandular-pubescent, ± spreading, deflexed towards the base; pedicels 10–25 mm long, spreading-pubescent, hairs of variable length, eglandular; bracts linear-lanceolate or rarely oblanceolate, 3–7(–16) mm long, pubescent. Calyx lobes (linear-)lanceolate, 2–3.5(–6) mm long, eglandular-pubescent, parallel to the corolla tube or spreading, the tips sometimes slightly recurved. Corolla pale to mid blue-violet, often paler on the palate, (20–)25–37 mm long, eglandular-pubescent outside; tube 12–25(–30) mm long, cylindric to the mid-point, where 1.5–3 mm deep and often slightly constricted, the floor then strongly downcurved to the mouth, glabrous within; mouth compressed laterally, inverted-V-shaped; limb bilabiate, spreading; upper lip of two erect, rounded lobes, 4–7 mm long, 5–9 mm wide, somewhat divaricate; lower lip 12–16 mm long, lateral lobes 4.5–9 mm long, 5.5–9.5 mm wide, median lobe 5–9 mm long, 5.5–10 mm wide, all rounded. Stamens

© The Board of Trustees of the Royal Botanic Gardens, Kew, 2006

arising from above the centre of the corolla tube; filaments 2–3 mm long, with stalked glands towards the apex; anther thecae rounded, 0.75–1.1 mm wide; staminodes minute. Ovary cylindric, 5–9 mm long, densely eglandular-pubescent; style 1.5–2(–3.5) mm long, eglandular-pubescent; stigma bilobed, 0.75–0.9 mm wide, papillose. Capsule (25–)40–55(–70) mm long, 1.5–1.7 mm diameter, pubescent. Seeds 0.5–0.8 mm long, reticulate.

TANZANIA. Rungwe District: Middle Fishing Camp, Kiwira River, Jan. 1963, *Richards* 17614! & Kiejo Volcano, W face, May 1975, *Hepper et al.* 5386! & Rungwe Forest Reserve, S side, Feb. 1986, *Bidgood et al.* 87!
DISTR. **T** 7; southern Malawi and northwestern Mozambique
HAB. Epiphytic in wet forest or on wet rocks in shade; 1200–2300 m
USES. None recorded on herbarium specimens
CONSERVATION NOTES. This species has a large range, though with a disjunct distribution, being absent from northern and central Malawi and northern Mozambique. Within its range, it has been collected on numerous occasions, particularly in southern Malawi where it is said to be frequent (*fide* la Croix in Kew. Mag. 11: 86 (1994)). However, insufficient information exists on the population sizes at its known sites or of the threats to their habitat, thus this species is provisionally assessed as Data Deficient (DD)

SYN. *S. mahonii* Hook.f. in Bot. Mag. 128: t. 7857 (1902); Baker & C.B. Clarke in F.T.A. 4(2): 505 (1906); Gilli in Ann. Naturhist. Mus. Wien 77: 54 (1973). Type: seed from Africa (?Zomba, Malawi), coll. *Mahon*, fl. in hort. R.B.G. Kew (K!, holo.)
    S. rungwensis Engl. in E.J. 57: 217 (1921); Gilli in in Ann. Naturhist. Mus. Wien 77: 54 (1973). Type: Tanzania, Rungwe, *Stolz* 1040 (B†, holo.; BM!, K!, PRE, Z, iso.)
    S. rungwensis Engl. var. *latifolius* Engl. in E.J. 57: 218 (1921). Type: Tanzania, Rungwe District, Mbaka, Kyejo, Mbungu River, *Stolz* 1040a (B†, holo.; BM!, iso.)
    [*S. tubiflos sensu* Baker & C.B.Clarke in F.T.A. 4(2): 506 (1906) and *sensu* Gilli in Ann. Naturhist. Mus. Wien 77: 55 (1973), *non* C.B.Clarke]
    [*S. breviflos sensu* Baker & C.B.Clarke in F.T.A. 4(2): 506 (1906), *non* C.B.Clarke]

NOTE. This species is highly variable, particularly in terms of number of inflorescences, number of flowers per inflorescence, inflorescence indumentum and flower size. Specimens from the Rungwe area of Tanzania often have only 1–2 inflorescences per plant with only (3–)8–11 flowers which tend towards the highest end of the size range, and the hairs of the inflorescence are often short. Gilli (in Ann. Naturhist. Mus. Wien 77: 54 (1973)) maintained that these plants be recognised as a separate species (*S. rungwensis* Engl.). Plants from southern Malawi and Mozambique tend towards 3 or more inflorescences per petiolode, often with many flowers which are usually at the lower end of the size range and the hairs of the inflorescence are variable in length, with some longer, spreading hairs. However, Hilliard & Burtt (1971: 188) note that even within isolated subpopulations individuals can display high variability, for example the type of *S. mahonii* Hook.f. has linear-lanceolate calyx lobes to 6 mm long, but revisiting of the type location revealed that the length of the calyx varies here from 3–6 mm, and thus overlapping with the range of typical *S. goetzei* flowers. Similarly, in Tanzania, individuals with flowers towards the lower end of the size range (*Hepper, Field & Mhoro* 5386!) and plants with up to 40 flowers on the inflorescence (*Goetze* 854!) are also recorded, in contrast to the general trend. Therefore, although such geographic variability exists, it is not considered sufficiently uniform to warrant subspecific recognition. Furthermore, the plants from across the range are united by the characteristic corolla shape.
    A specimen from the Kanda Hills in Ufipa District, **T** 4 (*Richards* 11120!) matches this species in the short, eglandular indumentum, short calyx lobes, lilac flowers and slender young fruits. The single leaf is rather large and proportionally broader than those typical of the Rungwe populations (though similar to plants from Malawi). The corollas are all either withered or in bud thus further collections are required from this range of hills in order to confirm this identification.

32. **Streptocarpus kungwensis** *Hilliard & B.L.Burtt* in Notes Roy. Bot. Gard. Edinb. 28: 210 (1968); Hilliard & B.L. Burtt, Streptocarpus: 200, fig. 31 (1971). Type: Tanzania, Mpanda District, W slope of Musenabantu, coll. *Harley* 9370, cult. in R.B.G. Edinb., C.4673 (E!, holo.)

© The Board of Trustees of the Royal Botanic Gardens, Kew, 2006

Monocarpic acaulescent herb. Unifoliate, rarely plurifoliate in cultivation; blade 10–27(–40) cm long, 6–20 cm wide, apex usually withered, base cordate to shortly attenuate, margin (serrulate-)crenulate, surfaces ± densely pubescent, particularly on the nerves beneath; lateral nerves (10–)18–20 pairs, spreading, conspicuous beneath; petiolode to 1.5 cm long, pubescent. Inflorescences 1–6, erect, arising from the base of the leaf blade, subcorymbiform to somewhat elongate, 10–30-flowered; peduncles 5.5–15 cm long; pedicels 8–18(–25) mm long, both densely short-pubescent with or without some glandular hairs, the latter sometimes more dense on the pedicels; bracts oblanceolate, 6–11 mm long, pubescent. Calyx lobes linear-lanceolate, (3.3–)5–6.5 mm long, eglandular-pubescent, sometimes with glandular hairs towards the base. Corolla pale lilac outside, often whitish on the tube, lilac to purple on the limb within with a white palate and base of lobes of the lower lip and two dark purple blotches in the throat, 24–35 mm long, eglandular- or glandular-pubescent outside, glabrous within; tube somewhat drooping, 15–22 mm long, cylindric in the lower half where (2.5–)3–4 mm deep, expanding gradually in the upper half to (6.5–)8–10 mm deep at the open, (sub-)rounded mouth; limb bilabiate; upper lip of two rounded lobes, 5–8 mm long, 4.5–9.5 mm wide, divergent; lower lip 9.5–17 mm long, lobes rounded, each 5.5–9 mm long, 6–10 mm wide. Stamens arising midway to two-thirds along the corolla tube; filaments pale violet, curved, 2.5–4.5 mm long, with scattered sessile or short-stalked glands towards the apex; anther thecae rounded, 1–1.5 mm wide; staminodes to 0.7 mm long, arising below the stamens. Ovary cylindric, 7–12 mm long, densely short-pubescent, the hairs spreading, largely or exclusively eglandular; style white, (1.5–)2.5–4.5 mm long, pubescent; stigma bilobed, 0.8–1.3 mm wide. Capsule 55–62 mm long, 2–3 mm diameter, pubescent. Seeds 0.4–0.5 mm long, reticulate.

TANZANIA. Mpanda District: W slope of Musenabantu, Aug. 1959, *Harley* 9370!; Iringa District: Kigogo, Mufindi, Apr. 1972, *Paget-Wilkes* 1052! & Kigogo River, Mar. 1993, *de Leyser* 291!
DISTR. **T** 4, 7; not known elsewhere
HAB. Rock faces in riverine forest and amongst rocks in riverbeds; 1350–1850 m
USES. None recorded on herbarium specimens
CONSERVATION NOTES. This species is currently known from only two highly isolated localities, the first in the Mahali Mts near Lake Tanganyika in westernmost Tanzania, the second in the Kigogo area of Iringa District. Paget Wilkes recorded it as growing in large colonies at the latter site in 1972, but no observation of its abundance there was recorded by de Leyser in 1993. There are no records from the intervening highlands, despite several collecting trips in this region. The populations therefore appear small and isolated, and this species is provisionally assessed as Endangered (EN B2ab(iii)), though more information on the status of the known populations is desirable

NOTE. Hilliard & Burtt (1971: 201) state that this species is close to *S. eylesii* S.Moore from which it is separable partly on the basis of it lacking the glandular indumentum of that species. Although certainly more dense in the latter species, glandular-pubescence is, however, recorded in *S. kungwensis*, only plants from the type locality lacking glandular hairs on the pedicels and outside of the corolla. More significant is the fact that, with the exception of *Procter* 634! from Mt Kungwe, *S. kungwensis* has almost exclusively eglandular hairs on the ovary. This character, together with the smaller, unscented flowers and smaller stigma, is diagnostic. The *Procter* collection, placed under *S. kungwensis* by Hilliard & Burtt despite its rather dense glandular indumentum (including the ovary) and longer, more conspicuous eglandular hairs, is certainly atypical and may well represent a hybrid between this species and *S. eylesii*.

33. **Streptocarpus compressus** B.L.Burtt in Notes Roy. Bot. Gard. Edinb. 22: 570 (1958); Hilliard & B.L. Burtt, Streptocarpus: 188, t. 7, fig. 27 (1971). Type: Tanzania, Songea District, Matengo Hills, Liwiri-Kiteza, *Milne-Redhead & Taylor* 8982 (K!, holo.; E, iso.)

Perennial acaulescent herb. Unifoliate to plurifoliate; blade broadly ovate, 11–28 cm long, 7.5–22.5 cm wide, base shallowly cordate, margin obscurely crenate to crenate-dentate, apex always withered in the material seen, surfaces pubescent, hairs of

© The Board of Trustees of the Royal Botanic Gardens, Kew, 2006

variable length, most dense on the nerves beneath; lateral nerves 14–17 pairs, parallel, spreading, impressed above, sometimes purplish beneath; petiolode less than 1 cm long, pilose. Inflorescences 3(–6), erect, arising from the petiolode, elongate or subcorymbiform, 8–50-flowered; peduncles 3.5–20 cm long, pilose, sometimes sparsely so, with some shorter glandular hairs; pedicels 10–30(–40) mm long, pilose with mixed glandular and eglandular hairs; bracts linear, 4–4.5 mm long, pubescent. Calyx lobes lanceolate to linear-lanceolate, 2.5–6 mm long, eglandular-pubescent with some glandular hairs towards the base, spreading, often slightly recurved at the tips. Corolla pale- to mid-blue-violet with dark blotching on the paler palate, 22–33 mm long, glandular-pilose outside, glabrous to sparsely glandular-pilose towards the mouth within; tube 10–17 mm long, cylindric to the mid-point, where 2.5–3 mm deep and sometimes slightly constricted, the floor then strongly downcurved to the mouth, this strongly compressed laterally, slit-like or inverted-V-shaped; limb bilabiate; upper lip of two erect, rounded lobes, 5.5–7(–10) mm long, 5–7(–9) mm wide, divaricate; lower lip held downwards, 14–21 mm long, lateral lobes 7–9.5 mm long, 9–10.5 mm wide, median lobe 8.5–10 mm long, 10–11 mm wide, all rounded. Stamens arising from above the centre of the corolla tube; filaments 2.5–3 mm long, with stalked glands towards the apex; anther thecae rounded, ± 1.2 mm wide; staminodes minute. Ovary cylindric, 5–7.5 mm long, densely to sparsely glandular-pubescent or glabrous except for sessile glands; style 1–2 mm long, glandular-pubescent; stigma bilobed, 1–1.3 mm wide, papillose. Capsule 30–45(–50) mm long, 1.5–3 mm diameter, pubescent to glabrescent. Seeds 0.5–0.7 mm long, reticulate.

TANZANIA. Iringa District: Mt Image, N of the Morogoro road, Mar. 1962, *Polhill & Paulo* 1632!; Njombe District: Rudewa, May, year not recorded, *Spurrier* 702!; Songea District: Mbinga, Kiteza, Mar. 1987, *Congdon* 154!
DISTR. **T** 7, 8; not known elsewhere
HAB. Rock crevices in shaded or exposed locations, rarely epiphytic; 1800–2200 m
USES. None recorded on herbarium specimens
CONSERVATION NOTES. This species is currently known from only the four locations cited, at only one of which (Matengo Hills) it has been collected on more than one occasion. Herbarium collections of the species provide no indication as to its abundance at these sites, or of potential threats to the populations. The extent to which this species is affected by forest loss requires further investigation, it being most often recorded on rock faces even in exposed situations. It is therefore currently assessed as Data Deficient (DD) but will likely qualify as Vulnerable following further *in situ* investigation of the populations

NOTE. Significant variation is recorded within the material currently placed within this species. Specimens from Mt Image and Rudewa display rather densely glandular pilose inflorescences whilst those from the Matengo Hills are more variable, usually sparsely glandular-pilose or even (in the case of the type) with an absence of hairs on the ovary, although *Zimmer* s.n.! from the same area has a more dense glandular indumentum. The Mt Image specimen has the most strongly compressed corolla mouth, it being slit-like. The Rudewa specimen has pedicels (2–4 cm) and calyx lobes (4–6 mm) at the longest end of the size range, the other collections having pedicels to only 2 cm and calyx lobes to only 4 mm Due to the paucity of material currently available, the consistency of these differences cannot be ascertained. In light of such uncertainty, it is considered most prudent to maintain these collections within one variable species.

34. **Streptocarpus eylesii** *S.Moore* in J.B. 57: 245 (1919); Hilliard & B.L. Burtt, Streptocarpus: 193, t. 8, fig. 29 (1971); Cribb & Leedal, Mountain flow. south. Tanzania: 123, t. 30 (1982); Hilliard & B.L. Burtt in F.Z. 8(3): 48 (1988). Type: Zimbabwe, Matopo Hills, *Eyles* 1097 (BM!, holo.; PRE, SRGH, Z, iso.)

Monocarpic, unifoliate to perennial, plurifoliate acaulescent herb. Leaves often pendant; blade broadly ovate, oblong or obovate, (4–)8–30 cm long, (2.5–)5–20 cm wide, base shallowly cordate, truncate or shortly attenuate, margin crenate-dentate, sometimes irregularly so, apex usually withered, surfaces densely long-pubescent;

© The Board of Trustees of the Royal Botanic Gardens, Kew, 2006

lateral nerves 7–14(–24) pairs, parallel, spreading; petiolode usually less than 1 cm long but occasionally to 2 cm, densely pubescent. Inflorescences 1–4, erect, arising from the petiolode and/or base of the leaf midrib, panicles subcorymbiform to somewhat elongate, 4–30-flowered; peduncles (2.5–)8–18.5 cm long, pilose with eglandular and sometimes glandular hairs; pedicels 15–35 mm long, densely pilose, with or without glandular hairs; bracts linear to oblanceolate, 5–16 mm long, pubescent. Calyx lobes lanceolate to linear-lanceolate, 3–9.5 mm long, pilose outside, with or without glandular hairs, eglandular-pubescent within. Flowers (?sometimes) smelling of creosote and honey. Corolla pendulous, variously pale blue-violet to white, sometimes darker on the limb than the tube, with either a paler (white or yellow) or darker (purple) mouth, (32–)37–60 mm long, pilose or glandular-pilose outside; tube (16–)20–40 mm long, 3–6 mm deep in the centre, downcurved above the centre and expanding to the mouth, where 10–15 mm deep; limb strongly bilabiate; upper lip of two suberect rounded lobes (5.5–)7–12 mm long, (6–)8–13 mm wide; lower lip 15–28 mm long, lateral lobes 8–15 mm long, 8.5–17.5 mm wide, median lobe 10–17.5 mm long, (8–)10–19.5 mm wide, all rounded. Stamens arising above the middle of the corolla tube; filaments 4–6(–8) mm long, slightly thickened centrally, with stalked or subsessile glands; anther thecae rounded, 1.4–2 mm wide; staminodes 0.5–0.75 mm long. Ovary cylindric, 6–17.5 mm long, densely pubescent, hairs short, spreading, predominantly glandular; style 1.5–7 mm long, glandular-pubescent; stigma shallowly bilobed, 1–2.5 mm wide, papillose. Capsule 20–60(–70) mm long, 2–3.5 mm diameter, pubescent. Seeds 0.5–0.8 mm long, reticulate.

a. subsp. **brevistylus** *Hilliard & B.L.Burtt* in Notes Roy. Bot. Gard. Edinb. 28: 210 (1968) & in Streptocarpus: 194 (1971) & in F.Z. 8(3): 49 (1988). Type: Malawi, S Vipya, Wozi Hill near Chikangawa, *Hilliard & B.L. Burtt* 4204 (E!, holo.; MAL, NU, iso.)

Unifoliate to plurifoliate. Leaf blade to 30 cm long, 20 cm wide. Peduncles with mixed eglandular and glandular hairs; pedicels mainly glandular-pilose. Calyx lobes 5–9 mm long, hairs mainly eglandular outside. Corolla pale blue-violet or rarely white, limb usually darker within, throat white to deep violet, 37–60 mm long, glandular-pilose outside; tube 20–35 mm long; limb with upper lobes 7–12 mm long, 8–13 mm wide, lateral lobes 9–15 mm long, 9–17.5 mm wide, median lobe 10–17.5 mm long, 10–19.5 mm wide. Filaments 5–6 mm long. Ovary 6–13 mm long; style (1.5–)3–4(–4.7) mm long; stigma (1–)1.5–2(–2.5) mm wide. Capsule 40–60(–70) mm long. Fig. 5: 1–10, p. 45.

TANZANIA. Kigoma District: 158 km from Mpanda on the Uvinza road, Nov. 1962, *Verdcourt* 3430!; Mbeya District: Loleza Mt behind Mbeya, Feb. 1979, *Cribb et al.*11395!; Njombe District: Madunda Mission, Feb. 1961, *Richards* 14085!
DISTR. **T** 4, 7; northern Zambia, northern and central Malawi and (?) easternmost Congo-Kinshasa
HAB. In rock crevices or shaded banks at forest margins or grassy slopes; 1700–2200(–2900) m
USES. None recorded on herbarium specimens
CONSERVATION NOTES. This subspecies has a localised distribution, being restricted to the northern part of the species' range. It can however be locally frequent, and its principal habitat of rocky hillsides is largely unthreatened by human activity; thus it is assessed here as of Least Concern (LC)

NOTE. *S. eylesii* comprises four subspecies. Hilliard & B.L. Burtt (1971: 194) separated subsp. *brevistylus* from the typical subspecies, of central Malawi southwards, principally on the basis of the shorter ovary and style, the pistil extending beyond the downcurve in the corolla tube in subsp. *eylesii* but only reaching the point of downcurve in subsp. *brevistylus*. Although this is generally true, the measurements show some overlap, subsp. *eylesii* having an ovary 9.5–17.5 mm long (the upper limit of 24 mm noted by Hillliard & Burtt not being recorded here) and a style 3.5–7 mm long (again, less than the maximum previously noted). However, several other distinguishing characters aid separation of these subspecies, notably that the calyx lobes are lanceolate, not linear-lanceolate, in subsp. *eylesii*, where they often measure only 3.5–5 mm long (less commonly extending to the 5.5–7 mm lobes recorded by Hilliard & Burtt), that the lateral corolla lobes are held forward in subsp. *brevistylus* but are more spreading and patent to the tube in subsp. *eylesii*, and that the stigma is somewhat broader in subsp. *brevistylus* (though stigmas of 3 mm broad as recorded by Hilliard & Burtt were not

© The Board of Trustees of the Royal Botanic Gardens, Kew, 2006

FIG. 5. *STREPTOCARPUS EYLESII* subsp. *BREVISTYLUS* — **1**, habit × ¹/₃; **2**, calyx and corolla, view from above × 2; **3**, detail of external corolla indumentum; **4**, calyx and corolla, lateral view × 2; **5**, dissected corolla showing the position of the stamens and staminodes × 2; **6**, detail of stamens × 4; **7**, pistil within calyx × 2; **8**, pistil and disk, calyx removed × 2; **9**, mature capsule with persistent calyx × ¹/₃; **10**, seed × 33.; subsp. *CHALENSIS* — **11**, mature capsule with persistent calyx × ¹/₃. 1, 2, 3, 4, 5, 6, 9, 10 from *Cribb et al.* 11395; 7, 8 from *Cribb et al.* 10816; 11 from *Bidgood et al.* 3717. Drawn by Riziki Kateya.

© The Board of Trustees of the Royal Botanic Gardens, Kew, 2006

noted here). In addition, the inflorescences of subsp. *brevistylus* are always rather densely glandular-hairy, this sometimes being sparse in subsp. *eylesii*. Hilliard & Burtt suggest that specimens of subsp. *eylesii* with short calyx lobes may be a result of hybridisation with *S. solenanthus* but this seems unlikely in view of the fact that all other features of the inflorescences of these specimens agree closely with *S. eylesii*.

    S. eylesii has close affinity to *S. monophyllus* Welw., a poorly known species from western Angola. The two are separated by the latter having a broader and less downcurved corolla tube which is held in a more upright position. However, study of further herbarium material and ideally live specimens of *S. monophyllus* is required to fully delimit these taxa; *S. eylesii* may prove to only merit subspecific status but in light of the current uncertainty it is maintained as separate here.

    Cribb et al. 11379! from the Poroto Mts, Tanzania has an atypically long and slender corolla tube and pistil for the species as a whole and particularly for subsp. *brevistylus* but otherwise closely resembles other material of that taxon. Further material from that location is required to determine whether this is an aberrant specimen or represents a distinct entity.

    b. subsp. **chalensis** *I.Darbysh.* **subsp. nov.** subsp. brevistylo similis sed corolla breviore alba atque labio inferiore flavo-maculato (nec caeruleo-violaceo vel omnino albo), et capsula breviore differt. A subsp. *silvicola* foliolis pluribus, corolla breviore et habitatione dissimili differt. Type: Tanzania, Ufipa District, Chala Mt, *Bidgood, Sitoni, Vollesen & Whitehouse* 3717 (K!, holo.; C, NHT, iso.)

Plurifoliate. Leaf blade to 15 cm long, 8.5 cm wide. Peduncles with mixed eglandular and glandular hairs; pedicels mainly glandular-pilose. Calyx lobes 5.5–6 mm long, hairs mainly glandular outside. Corolla white with a yellow spot on the lower lip, ± 32 mm long, glandular-pilose outside; tube 16–18 mm long; limb with upper lobes ± 5.5 mm long, 6 mm wide, lateral lobes ± 8 mm long, 9 mm wide, median lobe damaged in the material seen. Filaments 4 mm long. Ovary 6.5 mm long; style 3.5 mm long; stigma 1.2 mm wide. Capsule 20–30 mm long. Fig. 5: 11, p. 45.

TANZANIA. Ufipa District: Chala Mt, May 1997, *Bidgood et al.* 3717! (type)
DISTR. **T** 4; known only from the type location
HAB. Rock crevices in montane grassland; 1900–2200 m
USES. None recorded on herbarium specimens
CONSERVATION NOTES. This subspecies is currently known only from the type collection, and as such appears restricted to Mt Chala, although the highlands of southwestern-most Tanzania remain under-explored botanically and this taxon may therefore prove to be more widespread. Its rocky montane grassland habitat is largely unthreatened by human impact, but stochastic events at the single known location, such as fire or localised drought, may cause a significant decline in population. This subspecies is therefore assessed as Vulnerable (VU D2)

NOTE. From the limited material available (only one good corolla seen), this subspecies appears remarkable in the small flowers and short fruits. Within the species, it is closest to subsp. *silvicola* Hilliard & B.L.Burtt of central and southern Malawi, sharing in particular the white flowers with yellow on the palate and throat. Subsp. *silvicola* also displays fruits towards the lower end of the size range for this species, often 35–40 mm long, and rarely within the range of subsp. *chalensis*. However, subsp. *silvicola* differs in being unifoliate, in having longer corollas, with a minimum length of 35 mm and often within the range of subsp. *brevistylus* (to 55 mm long, *fide* Hilliard & B.L. Burtt in F.Z. 8(3): 49 (1988)), and in being a plant of rocky *Brachystegia*-woodland at 1000–1350 m alt. More material of subsp. *chalensis* is desirable to confirm the size range of the corollas and fruits.

35. **Streptocarpus sp. D**

Acaulescent herb. Unifoliate or usually bifoliate; principal blade broadly ovate to oblong-elliptic, 8–25 cm long, 5.5–20 cm wide, base cordate to shortly attenuate, margin crenate-serrate, apex withered in all material seen, surfaces ± densely pubescent; lateral nerves 13–16 pairs, parallel, spreading; petiolode short and inconspicuous. Inflorescences 2–6, erect, arising from the base of the leaf blade, panicles somewhat elongate, 4–10-flowered or more; peduncles 2.5–7.5 cm long, with

© The Board of Trustees of the Royal Botanic Gardens, Kew, 2006

mixed longer eglandular and shorter glandular hairs; pedicels 8–20 mm long, indumentum as the peduncles but with more numerous glandular hairs; bracts linear, 5–9 mm long, pubescent. Calyx lobes linear-lanceolate, 3.8–5 mm long, pilose outside, with occasional glandular hairs. Corolla somewhat drooping, blue(-purple), the limb appearing darker than the tube in herbarium material, 29–34 mm long, densely glandular-pubescent and with occasional longer eglandular hairs outside, glabrous within; tube strongly curved, 20–25 mm long, 2.5–4 mm deep in the centre, expanding towards the mouth where 6–8.5 mm deep; limb bilabiate, the lobes all held forwards and not strongly spreading; upper lip of two rounded lobes, 4–4.5 mm long and wide; lower lip 9–11 mm long, lateral lobes 3.5–5.5 mm long, 4.5–5 mm wide, median lobe 3.5–6 mm long, 3.5–5.5 mm wide, all rounded. Stamens arising in the upper third of the corolla tube; filaments 4–6.5 mm long, with subsessile or short-stalked glands most dense towards the apex; anthers partially or wholly exserted from the corolla tube, thecae rounded, 1.2–1.7 mm wide; staminodes minute, arising below the stamens. Ovary cylindric, 5.5–9.5 mm long, densely pubescent, the hairs short, spreading, mixed glandular and eglandular; style 3–4 mm long, pubescent; stigma shallowly bilobed, 0.9–1.3(–1.8) mm wide, papillose. Capsule ± 60 mm long, 2–2.5 mm diameter, pubescent. Seeds ± 0.5 mm long, reticulate.

TANZANIA. Chunya District: Ntande Hill, Feb. 1974, *Bally & Carter* 16497A!; Mbeya District: Loteza Peak, Jan. 1978, *Leedal* 4869!
DISTR. **T** 7; not known elsewhere
HAB. In rock crevices on slopes with bushland; 1700–2150 m
USES. None recorded on herbarium specimens
CONSERVATION NOTES. This species is known only from the two specimens cited. It appears to favour rocky montane habitats which have little agricultural potential, thus threats from man may be minimal. However due to the highly restricted range, it is vulnerable to environmental effects such as prolonged droughts or fires, thus is likely to prove vulnerable (VU D2) once it is fully delimited

NOTE. Although this is likely a new species, I consider the material of this taxon currently available insufficient for formal description. It is close to the complex group of acaulescent species with a strongly curved and deepened corolla tube and a glandular indumentum to the inflorescence (in our region *S. kungwensis*, *S. compressus* and *S. eylesii*). It appears readily separable by the partial or complete exsertion of the anthers from the corolla tube, these being included within the tube in the other species. From the two herbarium collections seen (*Leedal* 4869 being a scant specimen), the five lobes to the corolla appear to be held forward, with a rather narrow sinus between each, perhaps in order to protect the anthers. Further collections (including material in alcohol) or studies of living plants are necessary to confirm this character and also to determine the shape of the corolla mouth, a significant character within this group of species.

The rather small corolla limb may lead to confusion with *S. mbeyensis* which also has stamens inserted in the upper third of the corolla tube, but it lacks the glandular indumentum of *S.* sp. D, has a more shallowly curved corolla tube and a much longer pistil.

EXCLUDED SPECIES

Streptocarpus zimmermannii *Engl.* in E.J. 57: 215 (1921); Hilliard & B.L. Burtt, Streptocarpus: 376 (1971). Types: Tanzania, Amani, *Zimmermann* 3347 (B†, syn.) & *Grote* 3484 (B†, syn.)

These specimens are believed to have been destroyed in Berlin during World War II. Engler placed this taxon within the *S. caulescens* group of species, though he recorded it as having thick, subleathery, crenulate leaves, features which have not subsequently been recorded within this group.

© The Board of Trustees of the Royal Botanic Gardens, Kew, 2006

3. SCHIZOBOEA

(Fritsch) B.L.Burtt in Notes Roy. Bot. Gard. Edinb. 33: 266 (1974); Weber in Pl.
Syst. Evol. 134: 183–192 (1980); B.L. Burtt in Fl. Cameroun 27: 15 (1984); Troupin
in Fl. Rwanda, Spermatophytes 3: 498 (1985)

Roettlera sect. *Schizoboea* Fritsch in E. & P.Pf., Nachtr. I zum III-IV: 300 (1897)

Weak caulescent herbs. Stems somewhat fleshy, soon glabrescent. Leaves opposite-
decussate, petiolate. Inflorescences terminal but often overtopped by axillary vegetative
shoots and thus appearing lateral; main axis strongly condensed, few-flowered; bracts
inconspicuous, often hidden by the immature lateral vegetative buds. Calyx divided to
the base into five equal lobes. Corolla gamopetalous; tube cylindric; limb five-lobed, the
lobes subequal. Stamens arising from below the centre of the corolla tube, the two
anterior ones only fertile, two lateral staminodes present, the third, posterior one
absent; filaments ascending and convergent; anther thecae somewhat divergent, the two
anthers connate at the apex, cohering face to face, rarely free. Disk small, annular.
Ovary cylindric, unilocular, placentae intrusive, T-shaped; tapering into the style; stigma
capitate. Capsule narrowly cylindric, dehiscing loculicidally, eventually splitting into a
loose bundle of ten strands (the two valves each splitting into the midrib, two semi-
valves, and two halves of the intrusive placentae). Seeds fusiform.

A monotypic genus closely related to *Streptocarpus* subgen. *Streptocarpella*. Indeed, recent
molecular evidence (Möller & Cronk in Syst. Geog. Pl. 71(2): 545–555 (2001)) place it within that
subgenus, allied to the *S. elongatus* Engl. group (in E Africa represented by *S. gonjaënsis*, *S. thysanotus*
and *S. kimbozanus*). It is, however, maintained as a separate genus here on the basis that the
inflorescence is exclusively terminal; in *Streptocarpus* the inflorescences are always axillary or, in
some acaulescent species, borne on the hypocotyl-derived leaf stalk, although they can eventually
appear to surpress the terminal vegetative bud in some species. The fact that the capsule of
Schizoboea is not twisted was previously used to separate it from *Streptocarpus*; however, several species
of the latter genus are now known to have straight or near-straight capsules, though usually in
species with significantly shorter fruits than in *Schizoboea*. The taxonomic significance of the
eventual splitting of the capsule into up to ten strands in *Schizoboea* is questionable. In some species
of *Streptocarpus* (notably *S. exsertus* from our region), the two placentae split away from the valves
and divide into two, and the valves show some evidence of splitting along the midrib, thus it is
conceivable that the capsule would eventually split into ten strands as in *Schizoboea*.

Schizoboea kamerunensis (*Engl.*) *B.L.Burtt* in Notes Roy. Bot. Gard. Edinb. 33: 266
(1974) & in Fl. Cameroun 27: 15, t. 4: 1 (1984); Troupin in Fl. Rwanda, Spermatophytes
3: 498, fig. 151 (1985). Type: Cameroon, Barombi, Kumba, *Preuss* 951 (B†, holo.; BM!,
K!, iso.)

Decumbent caulescent herb to 30 cm tall, sometimes mat-forming. Stems rooting
along the lower creeping stems at and between the nodes, sparsely pubescent when
young, soon glabrescent, a few hairs persistent at the nodes. Leaf pairs subequal to
slightly unequal; blade ovate(-elliptic), (2–)4–7 cm long, (1.2–)3–6 cm wide, base
rounded to subcordate, ± asymmetric, margin subentire or with obscure blunt
serrations, apex acute to shortly acuminate, surfaces glabrous or sparsely pilose
particularly towards the base, hairs always sparsely present on the margins, lateral nerves
5–8 pairs, ascending; petiole 2–5 cm long, pubescent. Inflorescences 2–10-flowered;
peduncles very short, to 1 mm long; pedicels 4–6 mm long, sparsely pilose, the hairs
occasionally gland-tipped; bracts inconspicuous, narrowly elliptic to lanceolate, to 5 mm
long, pubescent. Calyx lobes linear, often somewhat curved, 4.5–7 mm long, pilose, the
hairs occasionally gland-tipped, venation parallel. Corolla white, held horizontally,
7.5–10 mm long, sparsely pilose outside, glabrous within; tube broadly cylindric,
6–7 mm long, 2.5–3.5(–5) mm diameter, floor slightly downcurved above the middle,
mouth open, rounded; limb of five rounded lobes, 1–2 mm long. Stamens arising from
below centre of corolla tube; filaments 1.5–2.5 mm long, glabrous; anthers reniform,

FIG. 6. *SCHIZOBOEA KAMERUNENSIS* — **1**, habit in fruit × ²/₃; **2**, habit in flower, small-leaved variant × ²/₃; **3**, detail of calyx and corolla, lateral view × 4; **4**, pistil within partially dissected calyx × 6; **5**, inflorescence with undehisced capsule × 1.5; **6**, dehisced capsule × 2; **7**, stylised cross section of capsule to show the division into 10 strands, with attached seeds × 16; **8**, seed × 60. 1 from *Poulsen* 688; 2, 4 from *Drummond & Hemsley* 1590; 3 from *Osmaston* 2162 with ref. to *Drummond & Hemsley* 1590; 5 from *Wrigley & Melville* 619; 6, 7 from *Raynal* 20501; 8 from *Reekmans* 8949. Drawn by Juliet Williamson.

© The Board of Trustees of the Royal Botanic Gardens, Kew, 2006

lobes ± 0.6 mm wide; staminodes minute. Ovary 3–5 mm long, glabrous, narrowing into the 1.5–2.5 mm style; stigma capitate, shallowly bilobed, 0.6–1.2 mm wide, minutely papillose. Capsule 15–30 mm long, 1–1.7 mm diameter, glabrous, style persistent, dehisced strands straw-like in texture and colour. Seeds 0.3–0.4 mm long, longitudinally ridged with transverse ridges, these sometimes verruculose. Fig. 6, p. 49.

UGANDA. Toro District: Ruwenzori, Nyamwamba Valley, Jan. 1935, *Taylor* 3137! & R. Ruate, tributary of R. Nyamagasani, Aug. 1952, *Osmaston* 2162!; Ankole District: Bushenyi, Kashoya-Kitomi Forest Reserve, N of Nzozia R., Nov. 1994, *Poulsen et al.* 688!
TANZANIA. Morogoro District: Uluguru Mts, Mgeta R., gorge below Hululu falls, Mar. 1953, *Drummond & Hemsley* 1590!; Iringa District: Udzungwa Mt National Park, Oct. 2002, *Luke et al.* 9131!; Rungwe District: Rungwe, May 1912, *Stolz* 1238!
DISTR. **U** 2; **T** 6–7; highlands of central and eastern Africa from western Cameroon and Bioko to Congo-Kinshasa, Burundi, Rwanda, Uganda and Tanzania
HAB. Forest, particularly along rivers and in the splash zone of waterfalls, occasionally epiphytic, rarely in more open woodland; 1300–2050 m
USES. None recorded on herbarium specimens
CONSERVATION NOTES. Although widespread, this species is known from few collections and is thus likely scarce across its range. It may, however, be overlooked due to its small, inconspicuous inflorescences and its preference for rather inaccessible habitats such as deep forest and waterfall edges. It is therefore initially assessed here as Data Deficient (DD), with further information on abundance and potential threats required from across its range

SYN. *Didymocarpus kamerunensis* Engl. in E.J. 18: 79 (1893); Baker & C.B. Clarke in F.T.A. 4(2): 503 (1906); B.L. Burtt in F.W.T.A. ed. 2, 2: 382 (1963)
  Roettlera kamerunensis (Engl.) Fritsch in E. & P.Pf., Nachtr. I zum III-IV: 300 (1897)
  Didymocarpus bequaertii De Wild. in Rev. Zool. Afr. 8, Suppl. Bot.: 40 (1920). Types: Congo-Kinshasa, Mokoto-Masisi, *Bequaert* 6556 (BR, syn.) & Ruwenzori, *Bequaert* 3812 (BR, syn.) & Irumu, *Bequaert* 2956 (BR, K!, syn.)
  Didymocarpus stolzii Engl. in E.J. 57: 203 (1921). Type: Tanzania, Rungwe, *Stolz* 1238 (B†, holo.; K!, iso.)
  Didymocarpus stolzii Engl. var. *minor* Mansf. in N.B.G.B. 12: 94 (1934). Type: Tanzania, Morogoro District, Uluguru Mts, *Schlieben* 3421 (B†, holo.; BM!, BR, iso.)

NOTE. Specimens from the Uluguru and Udzungwa Mts display leaves at the smallest and most hairy extreme of this variable species and were previously treated as a separate variety (*Didymocarpus stolzii* var. *minor*). However, in view of the significant variation in these characters across its range, the view given by Burtt in the protologue of the genus *Schizoboea*, that such variation owes more to the species' susceptibility to the influence of environmental factors than to evolutionary trends, is followed here.

4. **SAINTPAULIA**

H.Wendl. in Gartenflora 42: 321, t. 1391 & fig. 66 (1893); Baker & C.B. Clarke in F.T.A. 4(2): 500 (1906); B.L. Burtt in Gard. Chron., ser. 3, 122: 22 (1947) & in Notes Roy. Bot. Gard. Edinb. 22: 555 (1958) & in Notes Roy. Bot. Gard. Edinb. 25: 191–195 (1964); D.R. Johansson in Biol. Conserv. 14: 45–62 (1978); Baatvik in Fragm. Fl. Geobot. Suppl. 2(1): 97–112 (1993); Möller & Cronk in Amer. Journ. Bot. 84: 956–965 (1997) & in Proc. Roy. Soc. Lond. B, 264: 1827–1836 (1997); Eastwood, Bytebier, H.Tye, A.Tye, Robertson & Maunder in Bot. Mag. 15: 49–62 (1998); G.P. Clarke in Bot. Mag. 15: 62–67 (1998); Lindqvist & Albert in K.B. 54: 363–377 (1999) & in Syst. Geogr. Pl. 71: 37–44 (2002); Watkins, Kolehmainen & Schulman, Wild Afr. Violet, Saintpaulia (Gesneriaceae): an interim guide (2002); Schulman & Kolehmainen in Scripta Bot. Belg. 29: 165–170 (2004)

Perennial herbs; rosulate or caulescent, then creeping. Stems often fleshy, producing adventitious roots, short and thick in rosulate plants and bearing prominent leaf scars below the current cluster of leaves, more slender and with distinct internodes in caulescent plants. Leaves petiolate, opposite or becoming alternate due to differential growth in some caulescent plants, pairs subequal; blade

© The Board of Trustees of the Royal Botanic Gardens, Kew, 2006

often fleshy, indumentum of the upper surface highly variable, inconspicuous sessile glands usually present, the lower surface pubescent particularly on the nerves. Inflorescences axillary, cymose, often reduced to one or two flowers; bracts small, linear to oblanceolate; bracteoles as bracts but smaller; pedicels often becoming contorted in fruit. Calyx divided almost to the base into five (linear-)lanceolate lobes with blunt, glandular apices. Corolla pale to deep (blue-)purple, rarely white or bicoloured, gamopetalous, pubescent outside principally on the limb, lobes ciliate with glandular and/or eglandular hairs, largely glabrous within; tube very short, cylindric; limb ± widely spreading, often patent to the tube, bilabiate; upper lip with two equal rounded to elliptic lobes; lower lip with a central palate and three divergent subequal lobes, the central one often somewhat broader than the lateral pair. Stamens arising from just below the mouth of the corolla tube, the anterior two fertile; two lateral staminodes present, often minute, the posterior staminode either present or absent, one or more of the staminodes rarely fertile in abnormal flowers; filaments straight or twisted, often upcurved, flattened, often strongly so particularly towards the base, glabrous; anthers exserted from the corolla tube, usually bright yellow, reniform, the two thecae confluent at dehiscence, the two anthers connate at the apex, cohering face to face. Disk annular. Ovary unilocular, short, usually densely appressed-pubescent or villose, rarely with only sessile glands; style held to one side of the stamens, slender, largely glabrous except towards the base where pubescent; stigma with a central depression or bilobed, papillose. Fruit a loculicidal capsule, ovoid to cylindric, straight, tardily dehiscent. Seeds numerous, small, fusiform, longitudinally ridged and most often verruculose.

A genus endemic to the Flora region, important in the horticultural industry where they are marketed under the name "African Violet", with over 2000 cultivars. Dr. B.L. Burtt carried out extensive taxonomic research into the genus during the 1940s–1960s, recognising 20 species with four further varieties. He based much of his work on material grown in cultivation at the Royal Botanic Gardens, Edinburgh and Kew from seeds and leaf cuttings sent by collectors in east Africa. Many of the taxa recognised were acknowledged to have very close affinity, with species delimitation particularly problematic in the east Usambara Mts. Much of the taxonomy was based upon vegetative characters, most notably habit and upper leaf surface indumentum. The flowers of the majority of taxa were found to be essentially inseparable morphologically, although corolla colour was used as a character of significance. As much of the widely distributed cultivated material was cloned from relatively few initial wild collections, the different taxa have been deemed readily separable and thus the taxonomy of the group has persisted.

With such a narrow species concept, it is inevitable that as further wild collections have become available populations of *Saintpaulia* have been found which do not fit easily into the current classification. Whilst this has led in some instances to the proposal of potentially new species, it has led some botanists to call for a broadening of the species concept within the genus, particularly where populations intermediate between previously discrete taxa have been discovered. Recent genetic work (for example Möller & Cronk in Proc. Roy. Soc. London, B 264: 1827–1836 (1997); Lindqvist & Albert in K.B. 54: 363–377 (1999) & in Syst. Geogr. Pl. 71: 37–44 (2002)) has further questioned the species delimitation. Having now obtained a coverage of all but one (*S. inconspicua* B.L.Burtt) of the formally described taxa, the majority of "species" are found to have poor genetic differentiation, leading Lindqvist & Albert (1999) to suggest that the Usambara and coastal populations (*S. ionantha* complex) may best be regarded as a single metapopulation rather than divided into discrete taxonomic entities. Under such a theory, it is the relative isolation of the different populations, rather than any significant morphologically or ecologically driven genetic barriers, which result in the large variation seen, but at least irregular interbreeding is likely continuing between these populations. This is supported by the fact that the majority of taxa cross readily in cultivation and produce fertile hybrids (Batcheller in Gloxinian 34(2): 6 (1984)). Furthermore, the fact that many of the cultivars have been produced from a single "species" (*S. ionantha* H.Wendl.) indicates the potentially large genetic variability within this taxon which may be borne out within different isolated populations under varying ecological conditions.

In light of the phylogenetic data, the taxonomic significance of the characters used in the classification of *Saintpaulia* has been questioned. Several, most notably leaf indumentum and pigmentation and flower colour, are postulated to be under the control of very few genes (Lindqvist & Albert in K.B. 54: 363–377 (1999)). The resultant variation seen may therefore represent polymorphism through genetic diversity within a single entity. Other characters are

© The Board of Trustees of the Royal Botanic Gardens, Kew, 2006

likely to be largely phenotypic in origin, for example habit. It is quite conceivable that those populations growing in lush, sheltered environments are more likely to develop long trailing stems whilst those in harsher environments such as vertical rock faces with periodic drought remain compact and rosulate. The fact that such variation persists in cultivation indicates that the populations have begun to evolve genetically to adapt to their microenvironment. However, this persistence is inconsistent, with some plants which are rosulate in the wild becoming trailing in cultivation and vice versa (*Simiyu*, pers. comm.). Similarly, the very fleshy leaves and petioles of some of the low altitude taxa persist in plants propagated in cultivation but are almost certainly of phenotypic origin. Other characters are of insufficient consistency to be applied in isolation. For example, capsule morphology is highly diverse in *Saintpaulia* and initially appears to be of taxonomic value. However, remarkable variation in this character is recorded even within individual populations. Within the *S. ionantha*-complex, there is a general lengthening and narrowing of the capsule moving from the coast westwards, but there is significant overlap between the different entities.

In light of the above discussion, the current study applies a broadened species concept and reduces the number of species of *Saintpaulia* from twenty to six. Of these, three (*S. inconspicua* B.L.Burtt, *S. pusilla* Engl. and *S. goetzeana* Engl.) are highly distinct, whilst the remaining three (*S. shumensis* B.L.Burtt, *S. ionantha* H.Wendl. and *S. teitensis* B.L.Burtt) form a complex which, taking a broad species concept, could be considered conspecific. However, the two most distinct entities are here separated at the species level from the *S. ionantha* group. One of these (*S. teitensis*) is isolated geographically, the single population being approximately 150 km from the nearest known population of *S. ionantha*. The second (*S. shumensis*) is isolated ecologically, being recorded from higher altitudes than geographically sympatric populations of *S. ionantha*. Within *S. ionantha*, ten entities, separable by compliments of characters which are not deemed to be of either sufficient significance or consistency to support species status, are recognised; there is a significant allopatric element to their distribution, thus the rank of subspecies is applied. In addition one variety, var. *diplotricha* (B.L.Burtt) I.Darbysh. is recognised within subsp. *ionantha*; this taxon is deemed to display an indumentum sufficiently distinct from the typical variety to warrant formal taxonomic recognition, populations being readily separable in the field (S. Simiyu, *pers. comm.*). Many of the other infraspecific taxa display significant variation between and within populations which could be given recognition at the rank of variety or form. However, a conscious decision has been taken not to do this within the current work. To consistently recognise all variations in, for example flower colour and size and leaf shape, would lead to the need for a very large number of taxonomic entities. It would seem irresponsible to create so many new names in this already complex group until a full and exhaustive revision has been carried out; the current treatment is seen as a first step in this process.

The revised taxonomy of *Saintpaulia* presented here largely agrees with the genetic data, with one significant difference. Lindqvist & Albert (2002) indicate that the *ionantha*-type *Saintpaulia* taxa from the Nguru Mts (*S. nitida* B.L.Burtt and *S. brevipilosa* B.L.Burtt) form a monophyletic group distinct from the *S. ionantha* complex. However, this is not supportable on morphological grounds; *S. nitida* (=*S. ionantha* subsp. *nitida*) proves one of the more difficult taxa to key out within the *ionantha* complex, particularly in dried material, and *S. brevipilosa* is found to display overlap with *S. ionantha* subsp. *velutina* to such an extent that it is reduced to synonymy within that taxon.

Genetic data (e.g. Moller & Cronk in Andrews et al. (eds.), Taxonomy Cult. Pl.: 253–264 (2000) & in Amer. Journ. Bot. 84: 956–965 (1997); Smith *et al.* in Edinb. J. Bot. 55: 1–11 (1998)) has consistently shown *Saintpaulia* to have evolved from within *Streptocarpus* subgen. *Streptocarpella*. This relationship is supported by morphological data, the two sharing a similar embryology and pollen type, and by the identical chromosome number (n = 15). Several morphological characters have been used to separate the two genera in the past. Authors have pointed to the principally rosulate habit of *Saintpaulia* as opposed to the caulescent habit of subgen. *Streptocarpella*. However, this is inconsistent as several taxa of *Saintpaulia* display caulescence. The use of fruit characters is also questionable: although *Saintpaulia* taxa consistently display untwisted fruits regardless of shape and size whilst those of *Streptocarpus* are usually spirally twisted, the occurrence of untwisted fruits in some species of the latter genus reduces the taxonomic value of this character. The only consistent difference between the genera therefore appears to be the very short corolla tube and the pollen-reward pollination mechanism in *Saintpaulia* as opposed to the long tube and nectar-reward mechanism in *Streptocarpus*. Such differences are likely a simple response to a shift in pollinator which may have been induced by the habitat specialisation of *Saintpaulia* (Moller & Cronk 2000). However, the resultant flower morphology in *Saintpaulia* is so distinct as to render it readily separable from *Streptocarpus*, thus *Saintpaulia* is here maintained as a distinct genus. Future revisional work may result in the reduction of *Saintpaulia* to a section within *Streptocarpus* subgen. *Streptocarpella*.

© The Board of Trustees of the Royal Botanic Gardens, Kew, 2006

1. Upper surface of leaves glabrous (the margin
   ciliate); calyx lobes recurved at the apex (**T** 6) ..   1. **S. inconspicua** (p. 53)
   Upper surface of leaves hairy throughout, often
   densely so; calyx lobes not recurved (though
   sometimes curved outwards in *S. pusilla*)* 2
2. Corolla white with the lobes of the upper lip
   strikingly darker, mauve, blue or violet (distinct
   even in dried material), limb 7–21 mm long; leaf
   blade 1.2–3.8(–4.5) cm long 3
   Corolla pale to deep blue or purple, sometimes
   with a darker eye but never with a conspicuously
   darker upper lip, rarely white throughout, limb
   (8–)15–33 mm long; leaf blade 1.7–12.5 cm long 4
3. Rosulate or shortly caulescent plants; capsule
   cylindric, 1.2–1.7 mm diameter; corolla limb
   subspreading, 7–14.5 mm long, the lobes with
   eglandular hairs on the margin; leaves (oblong-)
   ovate to elliptic, upper surface with sparse, long
   and conspicuous erect or arched hairs (**T** 6, 7) .   2. **S. pusilla** (p. 54)
   Trailing caulescent plants, stems up to 40 cm long;
   capsule broadly ovoid, 2.8–4.5 mm diameter;
   corolla limb widely spreading, 11–21 mm long,
   the lobes with predominantly glandular hairs on
   the margin; leaves orbicular to broadly ovate,
   upper surface more densely pilose (**T** 6) ....   3. **S. goetzeana** (p. 55)
4. Anther thecae 1–1.5 mm long; filaments slender;
   upper surface of leaves with rather sparse but
   conspicuous long erect to arched hairs, fig 7: 14
   (p. 58), blade 2–5 cm long (**T** 3, 6) ....   4. **S. shumensis** (p. 56)
   Anther thecae (1.3–)1.7–2.7 mm long; filaments
   more clearly flattened, broader; upper surface of
   leaves with a variable indumentum, if conspicuous
   long erect hairs present then these either more
   dense or interspersed with dense short hairs,
   rarely sparser (in *S. teitensis*) but then mature
   blade over 5 cm long 5
5. Mature leaves with the petiole attached at the cordate,
   rounded or obtuse base (**K** 7; **T** 3, 6, 7) ....   5. **S. ionantha** (p. 57)
   Mature leaves eventually subpeltate, the petiole
   attached towards the base of the leaf but clearly
   inserted away from the margin (young leaves
   usually with the petiole attached at the cordate
   base) (**K** 7) ....   6. **S. teitensis** (p. 71)

* Dr Elspeth Haston (E) has recently communicated the finding of a *Saintpaulia* from the
Uluguru Mts with glandular hairs on the leaves, a unique character in the genus. This may
prove to be an additional species which could readily be keyed out prior to couplet 2. Living
plants are currently being grown on at E but no fertile material is available.

   1. **Saintpaulia inconspicua** *B.L.Burtt* in Notes Roy. Bot. Gard. Edinb. 22: 557
(1958); Watkins *et al.*, Wild Afr. Violet: 31 (2002). Type: Tanzania, Morogoro District,
Uluguru Mts, Morogoro, Kisaki road, *E.M. Bruce* 328 (K!, holo.)

   Delicate caulescent herb. Stems weak, to 15 cm long but often much shorter,
appressed-pubescent. Leaves pale green beneath; blade (ovate-)elliptic, 1.8–5.5 cm
long, 1.3–3.2 cm wide, base ± asymmetric, rounded, obtuse or shortly attenuate,
margin irregularly undulate or subentire, apex obtuse to rounded, upper surface

© The Board of Trustees of the Royal Botanic Gardens, Kew, 2006

glabrous except along the margin where ciliate; lateral nerves 4–5 pairs, conspicuous beneath; petiole 0.7–5 cm long, appressed-pubescent. Inflorescences 1–6-flowered; peduncles 1.2–3.2 cm long; pedicels (2–)6–11(–16) mm long, both sparsely pilose; bracts linear-lanceolate, 1–2.5 mm long, appressed-pubescent. Calyx lobes linear-lanceolate, 2.3–3.7 mm long, apex shortly recurved, outer surface pubescent, the hairs ascending. Corolla white or ? pale blue, the throat sometimes blue-purple (see note), shortly pubescent outside principally on the limb, the marginal hairs eglandular; tube 2–2.6 mm long; limb ± spreading, ± 7–11 mm long; upper lip 3–4 mm long, divided almost to the base into 2 elliptic lobes, each 1.5–2 mm wide; lower lip 4–6.5 mm long, lobes oblong (-elliptic), 2.5–5 mm long, 2.7–3.5 mm wide. Filaments slender, 2–3 mm long; anther thecae 0.8–1.2 mm long; staminodes not seen. Ovary cylindric, ± 2.5 mm long, appressed-puberulous; style 1.2–2 mm long, largely glabrous; stigma 0.3–0.4 mm diameter, papillose. Capsule narrowly cylindric, 9.5–16.5 mm long, 1–1.5(–2) mm diameter, puberulous. Seeds 0.4–0.6 mm long, longitudinally ridged with scalariform transverse ridges, these somewhat verruculose.

TANZANIA. Morogoro District: Uluguru Mts, NNE valley of Magari, Dec. 1970, *Pócs et al.* 6296/J! & N Uluguru Forest Reserve, Vidago, Dec. 1993, *Kisena* 899! & Uluguru N Catchment Forest Reserve, Tegetero-Luhungo path, Jan. 2001, *Jannerup & Mhoro* 97!
DISTR. **T** 6; restricted to the northern Uluguru Mts
HAB. Moist (sub)montane forest, often on or under rocks; 1350–1800 m
USES. None recorded on herbarium specimens
CONSERVATION NOTES. With three of the seven collections seen of this taxon having been made within the last 15 years, this species can no longer be considered as probably extinct in the wild (as suggested in, for example, Watkins *et al.* 2002). It is, however, clearly rare and restricted to a small region of the Uluguru Mts where it may be threatened by continued forest loss particularly at the lower end of its altitudinal range. It is therefore assessed as Vulnerable (VU B1ab(iii)+2ab(iii))

NOTE. Flower colour is not well documented in this species and may be variable. *Jannerup & Mhoro* (97) note it as "white with bluish purple tube" which agrees with *Bruce's* (328) "white flowers with blue spot" and *Haston's* (106) "white with violet-blue throat". *Kisena* (898, 899) records "white" flowers whilst *Harris & Pócs* (4561) note them as "pale blue", although the single dried flower seen in the latter gathering appears rather two-toned with a paler limb.

2. **Saintpaulia pusilla** *Engl.* in E.J. 28: 481, t. 7 (1900); Baker & C.B. Clarke in F.T.A. 4(2): 501 (1906); B.L. Burtt in Notes Roy. Bot. Gard. Edinb. 22: 558 (1958); Watkins *et al.*, Wild Afr. Violet: 42 (2002). Type: Tanzania, Morogoro District, Uluguru Mts, Lukwangule Plateau, *Goetze* 205 (B†, holo.)

Delicate rosulate (or rarely shortly caulescent) trailing herb. Stems, if present, slender, pilose. Leaves often purplish or reddish beneath; blade variable, (oblong-) ovate to (oblong-)elliptic, 1.2–3.2(–4.5) cm long, 0.8–2(–2.7) cm wide, base ± asymmetric, rounded, subcordate or attenuate, margin entire or inconspicuously serrate, apex rounded to obtuse, upper surface with rather sparse long erect or arching hairs, the margin with more dense, shorter hairs; lateral nerves 3–4 pairs, inconspicuous; petiole 0.6–2.5(–6.5) cm long, ± densely pilose. Inflorescences 1–4-flowered; peduncles (0.7–)1.3–3.3 cm long; pedicels (2.5–)5–10(–17) mm long, both sparsely pilose, the hairs antrorse; bracts linear, 0.5–2 mm long, pilose. Calyx lobes lanceolate, 1.5–3 mm long in flower, apex sometimes slightly curved outwards in fruit, outer surface ascending-pilose. Corolla somewhat campanulate, white with the lobes of the upper lip mauve or blue, sparsely pilose on the limb outside and with shorter eglandular hairs along the margin of the lobes; tube 2–2.8 mm long; limb 7–14.5 mm long, ± spreading; upper lip 2.5–5 mm long, lobes elliptic or subrounded, 1.5–3.5 mm long, 1.5–3.5 mm wide; lower lip 4.5–9.5 mm long, lobes oblong(-elliptic), 3–6 mm long, 2–5.5 mm wide. Filaments flattened, 1.8–2.5 mm long; anther thecae 1–1.3 mm long; staminodes 3, to 0.5 mm long. Ovary cylindric, 1.3–1.7 mm long, appressed-pubescent, narrowed into the style, 1.3–2 mm long, largely glabrous;

© The Board of Trustees of the Royal Botanic Gardens, Kew, 2006

stigma asymmetrically bilobed, 0.3–0.7 mm diameter, papillose. Capsule cylindric, 7–11 mm long, 1.2–1.7 mm diameter, often somewhat asymmetric, sparsely pubescent or glabrescent, style persistent. Seeds 0.4–0.5 mm long, verruculose with longitudinal ridges.

TANZANIA. Kilosa District: Ukaguru Mts, road from Mandege to Uponela, 2 km N of Mandege, Aug. 1972, *Mabberley* 1362!; Morogoro District: Uluguru Mts, Bondwa Peak, Jan. 1953, *Eggeling* 6443! & Nguru Mts, S branch of Divue valley, 1 km W of Mlaguzi, Sept. 1989, *Pócs* 89224/E!
DISTR. T 6, 7; restricted to the Uluguru, Ukaguru, Nguru and Udzungwa Mts
HAB. Shaded rockfaces and mossy rocks in montane rainforest; 1300–1750 m
USES. None recorded on herbarium specimens
CONSERVATION NOTES. This is one of the most widespread *Saintpaulia* taxa, being recorded from four separate mountain ranges. However, it is highly localised, currently being known from only one site in both the Ukaguru and Nguru Mts. The absence of this species from several seemingly suitable sites may be due to highly specific ecological requirements which are not fully understood at present. The assessment of Data Deficient (DD) applied by Eastwood *et al.* (in Bot. Mag. 15: 59 (1998)) is thus maintained here. As the currently known sites are highly isolated, human impact upon it habitat is minimal, thus *S. pusilla* may prove to be unthreatened

NOTE. *Thulin & Mhoro* 3208! from Mt Mkambaku, Uluguru Mts is unusual in having a shortly caulescent habit; the leaves and petioles are at the largest end of the range, the hairs on the latter are very dense and subappressed when young and the seeds have transverse ridges rather than verrucae. The leaf indumentum, corolla size and capsule shape however clearly place it within *S. pusilla*.

3. **Saintpaulia goetzeana** *Engl.* in E.J. 28: 481, t. 6 (1900); Baker & C.B. Clarke in F.T.A. 4(2): 501 (1906); B.L. Burtt in Notes Roy. Bot. Gard. Edinb. 22: 562 (1958); Watkins *et al.*, Wild Afr. Violet: 28 (2002). Type: Tanzania, Morogoro District, Uluguru Mts, Lukwangule Plateau, *Goetze* 245 (B†, holo.)

Trailing caulescent herb. Stems to 40 cm long when mature, densely pilose. Leaves usually purplish or reddish beneath; blade orbicular to broadly ovate, 1.6–3.8 cm long, 1.4–3.6 cm wide, base often somewhat asymmetric, rounded to shallowly cordate, margin subentire or shallowly crenulate-serrulate, apex rounded to obtuse, surfaces densely pilose, particularly above, the hairs erect or curved; lateral nerves (3–)4 pairs, inconspicuous particularly above; petiole 0.7–5.5 cm long, densely pilose. Inflorescences 1–2(–5)-flowered; peduncles (1.5–)2–4 cm long; pedicels (3–)6–11 mm long, both pilose; bracts linear-lanceolate, 0.7–2(–2.5) mm long, pilose. Calyx lobes lanceolate, 2–3.7 mm long, outer surface densely pilose. Corolla white with the lobes of the upper lip blue to violet, pilose on the limb outside and with shorter, predominantly glandular hairs along the margin of the lobes; tube subcampanulate, 2–3 mm long; limb 11–21 mm long; upper lip 4–8 mm long, lobes rounded, 3–6.5 mm long, 4–6.5 mm wide; lower lip 7–12.5 mm long, lobes ± rounded, 4–8 mm long, 4–9.5 mm wide. Filaments rather slender, 2.5–4.5 mm long; anther thecae 1.2–2 mm long; staminodes 3, to 1 mm long. Ovary broadly ovoid, 1.3–1.7 mm long, villose; style 3.5–5 mm long; stigma 0.2–0.4 mm diameter, papillose. Capsule broadly ovoid, 5–9.5 mm long, 2.8–4.5 mm diameter, pilose, style persistent. Seeds 0.5–0.65 mm long, longitudinally ridged with scalariform transverse ridges, the latter sometimes verruculose.

TANZANIA. Morogoro District: Uluguru Mts, S Uluguru Forest Reserve, edge of Lukwangule Plateau, Mar. 1953, *Drummond & Hemsley* 3656! & SW ridge of Lupanga above Mbete village, Nov. 1970, *Pócs & Nchimbi* 6285/N! & N Uluguru Forest Reserve, Vidago, Dec. 1993, *Kisena* 910!
DISTR. T 6; restricted to the Uluguru Mts and ? Nguru Mts
HAB. Wet, mossy rocks in montane wet forest, rarely epiphytic; 1700–2150 m
USES. None recorded on herbarium specimens

© The Board of Trustees of the Royal Botanic Gardens, Kew, 2006

CONSERVATION NOTES. Although recorded from only a small range, *S. goetzeana* appears locally frequent in high altitude forest in the Uluguru Mts. Reported collections of this species from the Nguru Mts could not be confirmed as the material (*Baatvik* 86830 & 86831, UPS) was unavailable for study by the author. In view of this uncertainty, the conservation rating of Data Deficient (DD) applied by Eastwood *et al.* (in Bot. Mag. 15: 59 (1998)) is here maintained. However, as human disturbance is likely to be minimal in its high altutude sites, this species may currently be unthreatened

NOTE. Stem and leaf indumentum are somewhat variable in this species. The majority of collections have a very dense indumentum of rather long hairs. However, those from the southern end of the range, the Lukwangule Plateau (*Drummond & Hemsley* 3656) and the Mzinga River catchment (*Simiyu et al.* Mzinga 2b, 7b, 8c, 9b), have considerably shorter hairs though at a similar density; these collections also have proportionally the broadest and shortest fruits. *Michelmore* 861, from Lupanga Peak, has a notably lower density of hairs on the leaves than other collections from the same locality, particularly beneath, but is otherwise consistent with this distinctive taxon. Phylogenetic studies have also recorded notable variability within this species (Lindqvist & Albert in K.B. 54: 371 (1999)).

A specimen recently received from Dr Haston, collected from above Tegetero mission in the Uluguru Mts (*Haston* 99), is vegetatively similar to small specimens of *S. goetzeana* but it has minute flowers which lack darker upper lobes. The specimen seen is however scant and more ample material is needed for further analysis.

4. **Saintpaulia shumensis** B.L. *Burtt* in Notes Roy. Bot. Gard. Edinb. 21: 238 (1955) & in Notes Roy. Bot. Gard. Edinb. 22: 558 (1958); Iversen in Symb. Bot. Upsal. 28: 241 (1988); Watkins *et al.*, Wild Afr. Violet: 44 (2002). Type: Tanzania, Lushoto District, W Usambara Mts, Shume, World's View, coll. *Greenway* 7934, cult. in R.B.G. Kew (K, holo., missing; E!, iso.)

Compact rosulate herb. Stem stout, to 7 cm long but often much shorter and sometimes largely absent. Leaves dark green and shining above, pale green to whitish or sometimes purplish beneath; blade broadly ovate to orbicular, 2–5 cm long, 1.5–4.5 cm wide, base rounded to cordate, margin crenate(-serrate), apex rounded to broadly obtuse, upper surface with rather sparse long erect or arched hairs, the margin with more dense mixed long and short hairs; lateral nerves (4–)5–6 pairs, conspicuous beneath; petiole 1–7.5(–11) cm long, pilose and with shorter subappressed hairs. Inflorescences 1–4(–6)-flowered; peduncles 1–6 cm long; pedicels 7–19 mm long, both spreading- to antrorse-pilose and with numerous shorter hairs, these spreading to subappressed; bracts linear, 1–5 mm long, pilose. Calyx lobes lanceolate, 2–4.5 mm long in flower, extending to 2.5–6 mm in fruit, outer surface with ascending hairs of variable length. Corolla pale lilac to almost white with a darker purple spot towards the base of the upper lip or blue-purple throughout, shortly pubescent and with occasional longer hairs on the limb outside, with short glandular and/or eglandular hairs along the margin of the lobes; tube 1.5–3 mm long; limb (8–)15–25 mm long; upper lip (3.5–)5.5–9.5 mm long, lobes (rounded-)elliptic or somewhat obovate, (2.5–)4.5–7.5 mm long, (2–)4–7.5 mm wide; lower lip (4.5–)9–14.5 mm long, lobes (oblong-)elliptic or somewhat obovate, (3.5–)5.5–10.5 mm long, (3–)4.5–9.5 mm wide. Filaments slender, (2–)3–4.5 mm long; anther thecae 1–1.5 mm long; staminodes 3, minute. Ovary conical, 1.5–3 mm long, densely appressed-pubescent, narrowed into the style, 2.5–4.5 mm long; stigma ± bilobed, (0.3–)0.6–1.3 mm diameter, papillose. Capsule conical to cylindric, 7–16(–20) mm long, 1.2–2.5(–3.5) mm diameter, sometimes asymmetric, apex tapered, surface shortly pubescent, style ± persistent. Seeds 0.45–0.5 mm long, verruculose with longitudinal ridges. Fig. 7: 14 (leaf indumentum), p. 58.

TANZANIA. Lushoto District: Shume, World's View, Jan. 1953, *Eggeling* 6488! & Sungwi Forest Reserve, Aug. 1955, *Semsei* 2200!; Morogoro District: Nguru Mts, E edge of Mafulumla, W of Kombola village, *Pócs et al.* 87015/F!
DISTR. **T** 3, 6; restricted to the Usambara and Nguru Mts
HAB. Rock faces and rock crevices in montane forest; (?1300–)1500–2000 m

© The Board of Trustees of the Royal Botanic Gardens, Kew, 2006

USES. Cultivated as an ornamental

CONSERVATION NOTES. In the W Usambara Mts this species was recorded as "very common" at Sungwi (*Semsei* 2200) in the 1950s but its current status is unknown there. The type locality, World's View, Shume, has experienced significant loss of forest through conversion to agriculture and settlement, thus this population is likely to be severely depleted. Its status in the Nguru Mts is uncertain, but it is likely to experience only limited disturbance there due to its preference for high elevations. Its presence in the E Usambara Mts remains unconfirmed (see note). This species is thus assessed as Data Deficient (DD) at present but is likely to qualify as Vulnerable

SYN. [*S. pusilla sensu* Lindqvist & Albert in K.B. 54: 371 (1999), pro parte quoad *Pócs et al.* 87015/F, *non* Engl.]

NOTE. Dr. Burtt (1958) first recognised the possible occurrence of this species in the Nguru Mts, based upon a specimen from Kinbola (*Schlieben* 4112, B†) but suggested that the collector's failure to note the flowers as bicoloured negated against their affinity. However, uniformly coloured flowers have since been recorded in the Usambara populations of *S. shumensis* (Sungwi Forest Reserve, *Semsei* 2200!). Several additional specimens have become available from the Nguru Mts, but unfortunately mature flowers are absent. Several differences occur in the Nguru material, including the leaves sometimes being purplish beneath, not green-white, the shorter calyx lobes and filaments and the less conspicuously bilobed stigma. This latter character is a particularly notable feature of the Shume plants, although its taxonomic significance is currently unclear. Pending further analysis of living material, the Usambara and Nguru plants are here treated as conspecific. A fruiting specimen from the E Usambara Mts, labelled only "Tanga" (*Skarpe* 55!) is almost certainly of this species; further details on the collecting locality of this population are required. It is perhaps from the Nilo Peak, Lutindi Forest Reserve, where *S. shumensis* has been recorded from a sterile specimen (*Iversen et al.* 87471!). The leaves of the E Usambara material (particularly in the latter collection) are however rather atypical in being very shallowly crenate or subentire when mature. Flowering material is therefore required to confirm the placement of these populations within *S. shumensis*.

S. *shumensis* appears rather intermediate between the smaller Uluguru-Nguru *Saintpaulia* taxa, most notably *S. pusilla* (small plants from the Ngurus often mistaken for that taxon) and the *S. ionantha* complex (*S. ionantha* subsp. *velutina* being particularly close). In view of the postulated high elevation ancestry and evolution of the genus (Lindqvist & Albert in Syst. Geogr. Pl. 71: 42–43 (2002)), it is quite possible that this montane species provides the link between the basal taxa and the *S. ionantha* complex.

5. **Saintpaulia ionantha** *H.Wendl.* in Gartenflora 42: 321, t. 1391 (1893); Baillon in Bull. Soc. Linn. Paris 2: 1148 (1893); Hook. f. in Bot. Mag. 121, t. 7408 (1895); B.L. Burtt in Notes Roy. Bot. Gard. Edinb. 22: 559 (1958); Troupin in Fl. Rwanda, Spermatophytes 3: 496 (1985); Watkins *et al.*, Wild Afr. Violet: 33 (2002). Types: Tanzania, "about one hour from Tanga", coll. *von Saint-Paul-Illaire*, cult. in Herrenhausen, Hannover (illus. Gartenflora 42: 321) & cult in R.B.G. Kew (K!, epi., selected here; see note)

Rosulate to creeping caulescent herb. Stem variable, robust and short in rosulate plants when often decumbent, more slender and trailing for up to 20 cm or more in caulescent plants, sometimes branched, appressed-pubescent to -pilose particularly when young. Leaves varying from fleshy to rather thin, pale green to red-purple beneath; blade variable, ovate, (oblong-)elliptic or orbicular, 1.7–12.5 cm long, 1.5–9.5 cm wide, base cordate, rounded or obtuse, sometimes asymmetic, margin shallowly to coarsely crenate, (crenate-)serrate or subentire, apex rounded to acute or subattenuate, upper surface with a highly variable indumentum (see subspecies); lateral nerves 4–7 pairs, ± conspicuous beneath; petiole (1–)3–15(–24) cm long, with long and/or short hairs, variously spreading to appressed. Inflorescences 1–12-flowered; peduncles 1.5–10(–15) cm long; pedicels 7–36 mm long, both with an indumentum as the petioles; bracts linear to oblanceolate, 1–7.5 mm long, puberulous to pilose. Calyx lobes lanceolate,

© The Board of Trustees of the Royal Botanic Gardens, Kew, 2006

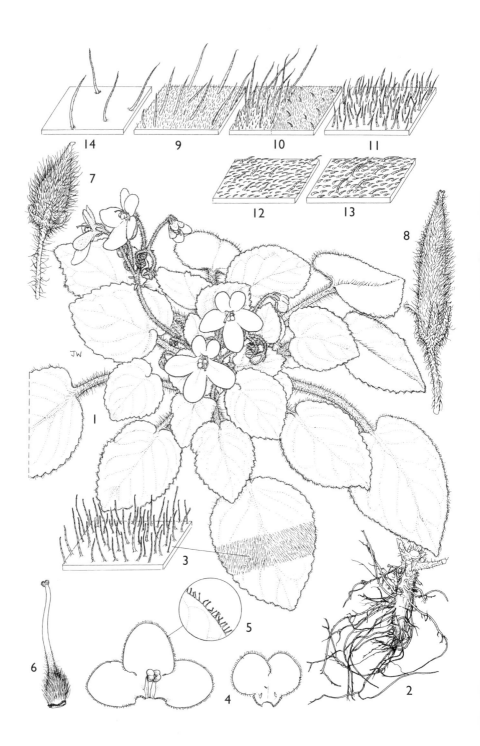

© The Board of Trustees of the Royal Botanic Gardens, Kew, 2006

1.5–4(–6) mm long in flower, extending somewhat in fruit, outer surface puberulous to pilose, the hairs subappressed to spreading. Corolla colour variable (see subspecies), pubescent to pilose on the limb outside, with short hairs along the margin of the lobes, these variously predominantly glandular, predominantly eglandular or mixed; tube (1.2–)1.5–3 mm long; limb 12–33 mm long; upper lip 5–12 mm long, lobes (oblong-)rounded, 3–9 mm long and wide; lower lip 6.5–19.5 mm long, lobes (oblong-) rounded to somewhat obovate, 4–12.5 mm long, 4–14 mm wide. Filaments flattened, (2–)2.5–3.5(–4) mm long; anther thecae (1.3–)1.7–2.7 mm long; staminodes 2–3, minute. Ovary ovoid to conical, 1–3.5 mm long, appressed-pubescent to villose; style 3–7.5 mm long; stigma shallowly bilobed, 0.3–0.6(–0.9) mm diameter, papillose. Capsule broadly ovoid to narrowly cylindric, 7–30 mm long, 1.5–5 mm diameter, surface sparsely to densely pubescent, style ± persistent. Seeds 0.4–0.65 mm long, verruculose with longitudinal ridges. Fig. 7: 1–13, p. 58.

NOTE. This highly complex and variable taxon is here divided into nine subspecies. They are defined by differing compliments of a set of highly variable characters, no single character being of use in isolation across the whole taxon. They can be difficult to distinguish, particularly in dried material where leaf hair orientation can be unclear (in such cases the orientation of the hairs on the petiole is often, though not always, a good guide). The mature leaves should be observed with regard to indumentum as the young leaves can be atypical in hair length and density. It should also be noted that some populations within several of the subspecies display a dual indumentum of both long and short hairs (**diplotrichous**). In such cases, the orientation of both hair types must be observed (and if making collections it is imperative to record this information on the data label). Where capsule shape and size is important, the largest capsules should be examined as some taxa display significant lengthening of the capsule with maturity. It is recommended that geography be used in association with the character key as the majority of subspecies have discrete geographic distributions; these are therefore listed in the key. It should also be noted that intermediate specimens occur, and are not uncommon in areas of range overlap.

The original description of *S. ionantha* was made from living cultivated material grown on from seed despatched by Baron von Saint-Paul-Illaire; no dried type specimen appears to have been preserved. The colour plate opposite page 321 in the protologue is thus applicable as the type. However, there is a slight confusion over the original despatch of seed, it having been taken from two separate locations (and representing two of the currently recognised subspecies, subsp. *ionantha* and subsp. *grotei*). Although the original description is clearly based largely on the material from near Tanga, the colour plate is not wholly diagnostic; it is thus complimented here by designation of an epitype. A specimen of material grown at Kew from the seed of plants derived from von Saint-Paul-Illaire's original Tanga material is here selected; this specimen was later used as the type of the superfluous name *S. kewensis* by C.B. Clarke.

FIG. 7. *SAINTPAULIA IONANTHA* subsp. *IONANTHA* var. *IONANTHA* — **1**, habit × ²/₃; **2**, root system and lower stem × ²/₃; **3**, detail of upper leaf surface indumentum × 4; **4**, dissected corolla, upper and lower lips with associated tube sections × 1; **5**, detail of corolla margin × 8; **6**, pistil × 3; **7**, capsule, typical ovoid form × 3; **8**, capsule, subcylindric variant × 3; **9–14** detail of upper leaf surface indumentum × 4: subsp. *IONANTHA* var. *DIPLOTRICHA* — **9**; subsp. *VELUTINA* — **10**, diplotrichous and puberulent variants; subsp. *PENDULA* — **11**; subsp. *GROTEI* — **12**; subsp. *NITIDA* — **13**; *SAINTPAULIA SHUMENSIS* — **14**. 1, 3 from cult. Kew 1987-1365; 2 from cult. Edinb. C.3789; 4, 5, 6 from cult. Kew s.n.; 7 from *Frontier-Tanzania* 306; 8 from *Herring* in *Moreau* 4; 9 from *Milne-Redhead & Taylor* 7287; 10 from cult. Kew 1974-2880 (left) and from cult. Kew 1995-503 (right); 11 from cult. Kew, 1982-5708; 12 from cult. Kew 1989-1527; 13 from cult. Kew 1997-4406; 14 from cult. Kew EN 399/1949. Drawn by Juliet Williamson.

© The Board of Trustees of the Royal Botanic Gardens, Kew, 2006

KEY TO INFRASPECIFIC VARIANTS

1. Upper surface of mature leaves with very
    sparse, short hairs only (**T** 3, Mafi Hill) ..   i. subsp. *mafiensis* (p. 70)
    Upper surface of mature leaves more densely
    hairy, although the hairs sometimes incon-
    spicuous (check with a hand lens) 2
2. Upper surface of leaves with all hairs erect or
    nearly so, either ± densely pilose or
    puberulent, the latter with or without sparse
    but conspicuous interspersed long hairs 3
    Upper surface of leaves with at least the
    shorter hairs appressed, the hairs often of a
    variable length (though rarely as extreme
    as in the diplotrichous taxa of the first lead
    of the couplet), then the longer ones either
    appressed, arced or subspreading 8
3. Upper leaf surface pilose, hairs of ± equal
    length (sometimes with occasional longer
    hairs particularly towards the margin), or if
    hairs shorter then leaf blade ovate to elliptic 4
    Upper leaf surface puberulent, or with a dual
    indumentum of sparse but conspicuous
    long hairs interspersed between dense
    puberulent hairs (diplotrichous); if long
    hairs absent then leaf blade broadly ovate
    to orbicular and upper surface with a
    velvet-like texture 7
4. Capsules narrowly cylindric, 12.5–30 mm
    long, the length 6–14.5 times that of the
    width; leaves either shortly pubescent or
    pilose, blade ovate to elliptic, the central
    ones of the rosette tending to be held
    rather upright; plants of exposed or
    shaded limestone ledges (**K** 7) ....   c. subsp. *rupicola* (p. 64)
    Capsules broadly ovoid to subcylindric, 7–18
    mm long, the length 1.8–6 times that of
    the width; leaves pilose, blade variously
    ovate, elliptic or suborbicular, the rosette
    tending to be more spreading (more rarely
    upright); plants of coastal to submontane
    forest, usually in wetter microhabitats 5
5. Plants often caulescent with distinct
    internodes between the lower leaves, more
    rarely rosulate; leaves ovate to suborbicular,
    3–6(–8.5) cm long, apex acute to
    subattenuate, more rarely obtuse; hairs on
    upper leaf surface erect or curved towards
    the apex (**T** 3, E Usambara Mts) ....   b. subsp. *pendula* (p. 63)
    Plants usually rosulate, if shortly caulescent
    then leaves elliptic; leaves suborbicular,
    ovate or (oblong-)elliptic, 3–10 cm long,
    apex rounded to obtuse or if acute then
    leaves (ovate-)elliptic; hairs on upper leaf
    surface erect 6

© The Board of Trustees of the Royal Botanic Gardens, Kew, 2006

6. Capsule ovoid, more rarely subcylindric, the
   length 1.7–5(–6) times that of the width; leaf
   and flower colour variable but if leaves pale
   green beneath, corolla usually pale mauve   a. subsp. *ionantha*
   (**T** 3, 6, 7, coastal lowlands & Udzungwa Mts)   var. *ionantha* (p. 62)
   Capsule shortly (sub-)cylindric, the length
   3–6 times that of the width; leaves always
   pale green beneath; corolla always deep
   (blue-)violet (**T** 3, W Usambara Mts) . . . .   d. subsp. *grandifolia* (p. 65)
7. Leaf surface with a velvet-like texture above,
   with or without interspersed long hairs,
   margin subentire or crenate(-serrate);
   capsule cylindric, the length 3.5–10 times
   the width (**T** 3, 6, W Usambara, Nguru
   and Uluguru Mts) . . . .   g. subsp. *velutina* (p. 68)
   Leaves not velvet-like above, always with
   interspersed long hairs, margin crenate-
   serrate; capsule ovoid to shortly
   subcylindric, the length 2–4.5 times the
   width (**T** 3, coastal lowlands and foothills of   a. subsp. *ionantha*
   E Usambara Mts) . . . .   var. *diplotricha* (p. 63)
8. Plants caulescent, the internodes rather
   densely pubescent; leaves shiny above; calyx
   lobes 3.8–5 mm long in flower; capsule
   cylindric, 18–25 mm long, the length 8–8.5
   times the width (**T** 3, W Usambara Mts) . .   j. subsp. *occidentalis* (p. 70)
   Plants rosulate or if caulescent then
   internodes sparsely pubescent, leaves rather
   dull above, calyx lobes 2.3–4.3 mm long and
   capsule conical to cylindric, 8–20 mm long,
   the length (2.5–)3.5–8(–10) times the width
   (note: these characters more variable in
   strictly rosulate taxa) 9
9. Upper leaf surface shiny, often dark green,
   densely appressed-puberulous, the hairs ±
   equal in length, leaf margin subentire to
   obscurely crenate when mature; corolla
   always with predominantly eglandular hairs
   on the margin of the lobes (**T** 6, Nguru Mts)   h. subsp. *nitida* (p. 69)
   Upper leaf surface dull or if shiny then paler
   green, the hairs less dense and of ± variable
   length; leaf margin coarsely to shallowly
   (crenate-)serrate; hairs on the margin of
   the corolla lobes variable 10
10. Inflorescence (2–)4–12-flowered; corolla
    with predominantly glandular hairs on the
    margin of the lobes; capsule ± narrowly
    cylindric, 11–25.5 mm long, the length
    (4.5–)7.5–13.5(–17) times the width (**T** 3,
    W Usambara Mts) . . . .   f. subsp. *orbicularis* (p. 67)
    Inflorescence 1–4(–6)-flowered; corolla with
    either predominantly eglandular or mixed
    glandular and eglandular hairs; capsule
    conical to cylindric, 8–20 mm long, the
    length (2.5–)3.5–8(–10) times the width
    (**T** 3, E Usambara Mts) . . . .   e. subsp. *grotei* (p. 66)

© The Board of Trustees of the Royal Botanic Gardens, Kew, 2006

a. subsp. **ionantha**

Plants rosulate or rarely shortly caulescent, fleshy throughout. Leaves often tinged purple beneath; blade ovate to (oblong-)elliptic, 3–10 cm long, margin obscurely to conspicuously crenate-serrate, apex rounded, obtuse or more rarely acute, upper surface with dense erect long hairs or diplotrichous. Petioles, peduncles and pedicels pilose, the hairs usually spreading, more rarely interspersed with ascending shorter hairs. Corolla pale mauve to deep blue-purple, rarely white, margin with predominantly glandular hairs. Capsule 7–15 mm long, 2.5–5 mm diameter.

i. var. **ionantha**

Leaf margin shallowly and often inconspicuously (crenate-)serrate, upper surface ± densely pilose. Petiole, peduncle and pedicel hairs spreading.

TANZANIA. Tanga District: Amboni Caves, Jan. 1992, *Raistrick* SP/8!; Handeni District: Gendagenda North Forest, Jan. 1992, *Raistrick* SP/9!; Rufiji District: Kiwengoma Forest, Nov. 1989, *Frontier-Tanzania* 252!
DISTR. **T** 3, 6, 7; restricted to the coastal lowlands and lower Udzungwa Mts
HAB. Shaded, usually moist, ledges and crevices in limestone or gneissic rocks, often in semi-deciduous forest; 0–750 m
USES. This taxon forms the basis of the cultivated "African Violet" industry
CONSERVATION NOTES. Under the broadened circumscription of this taxon presented here, it has been collected from at least 12 sites over a broad distribution. However, the lowland sites around Tanga are known to be severely threatened by habitat degradation and it is likely that several have been lost or will be so in the near future. Populations from the Udzungwa Mts appear less threatened whilst the status of the Kiwengoma population is unconfirmed. This taxon (and the species as a whole) is provisionally assessed as Vulnerable (VU A2c) as it is estimated that more than 30% of the population has been lost in the recent past and that population decline is continuing

SYN. *Petrocosmea ionantha* (H.Wendl.) Rodigas in Ill. Hort. 42: 108 (1895)
    Saintpaulia kewensis C.B. Clarke in F.T.A. 4(2): 501 (1906). Type: Tanzania, coll. *von Saint-Paul-Illaire*, cult in R.B.G. Kew (K!, holo.)
    S. tongwensis B.L. Burtt in Gard. Chron., ser. 3, 122: 23 (1947) & in Bot. Mag. 165: t. 11 (1948) & in Notes Roy. Bot. Gard. Edinb. 22: 560 (1958); Iversen in Symb. Bot. Upsal. 28: 243 (1988); G.P. Clarke in Bot. Mag. 15: 66 (1998); Watkins *et al.*, Wild Afr. Violet: 46 (2002). Type: Tanzania, Pangani District, Tongwe Mt, coll. *Moreau*, cult. in R.B.G. Kew (K!, holo.) **syn. nov.**
    [*S. intermedia sensu* D.R.Johansson in Biol. Cons. 14: 54–57 (1978) quoad spec. ex Sigi Falls]
    [*S.* "white *ionantha*" *sensu* J. Smith in Gloxinian 48(2): 44 (1998)]
    [*S.* sp. nov. "Sigi Falls" *sensu* Eastwood *et al.* in Bot. Mag. 15: 59 (1998); *S.* "Sigi Falls" *sensu* Watkins *et al.*, Wild Afr. Violet: 51 (2002); *S.* sp. nov. "Sigi Falls" *sensu* Schulman & Kolehmainen in Scripta Bot. Belg. 29: 168 (2004)]
    [*S.* sp. nov. "Pangani Falls" *sensu* Schulman & Kolehmainen in Scripta Bot. Belg. 29: 168 (2004)]

NOTE. *S. tongwensis* was separated from *S. ionantha* on the basis of having longer fruits, paler flowers and longer, elliptic leaves with a more acute apex. Fruit length however varies considerably within this taxon and, although plants from the Tanga region assigned to *S. ionantha* often display the most broadly ovoid fruits, there is considerable overlap. The overall range in length:width ratio of fruits from plants assigned to *S. tongwensis* (2.4–4.5(–6):1) is barely different to that of those assigned to *S. ionantha* (1.8–3.5(–5):1). Although most commonly blue-purple, *S. ionantha* is known to display significant variability in flower colour (see for example Burtt, 1958: 560), including white-flowered variants. The pale violet flowers of *S. tongwensis* clearly fall within this range. Leaf shape is initially the most striking difference between the two taxa, largely because the most well known populations of *S. ionantha* from the Tanga region usually have broadly ovate leaves. However, plants from further south in the Kiwengoma Forest, otherwise largely identical to the Tanga material, have more elongate leaves approaching *S. tongwensis*.
    Plants from Sigi Falls near Tanga have elongate, elliptic leaves with notably paler lateral nerves above. The flowers are rather deep blue-purple. This population was assigned to *S. intermedia* by Johannson, largely on account of the shortly caulescent habit of the wild plants.

© The Board of Trustees of the Royal Botanic Gardens, Kew, 2006

However plants from this population cultivated at Kew are rosulate in habit, thus the habit of the wild plants is likely a phenotypic character. With the elongate leaves and blue-purple flowers, it appears to be intermediate between *S. ionantha* and *S. tongwensis*; all three entities are here considered to belong to a single variable taxon.

Plants from the Udzungwa Mts (**T** 7), at or near Sanje Falls, growing in lowland rainforest appear very close to this taxon. Photographs of fruiting material recently collected there (W.R.Q. Luke, *pers. comm.*) clearly place it within the subsp. *ionantha* - subsp. *pendula* group, the fruits being squat and ovoid. The first collection from this region (Sanje Falls, *Polhill & Lovett* 5130!) has spreading hairs on the leaves and petioles and is largely inseparable from subsp. *ionantha*. However, a later collection (Sanje–Mwanihana route, *P.A. & W.R.Q. Luke* 5004!) has ascending hairs on the petiole, thus likely having similarly angled hairs on the upper leaf surface. This would tend to place that specimen closer to subsp. *pendula*, yet the obscurely serrate margin and obtuse apex to the leaves are consistent with subsp. *ionantha*. This material is therefore provisionally placed in the latter but requires further investigation. The cultivated material at R.B.G. Kew labelled as originating from the *Polhill & Lovett* collection (accession number 1983-8132) is clearly different to the herbarium specimen of the wild material and is believed to have been mislabelled (it is referable to subsp. *pendula*).

ii. var. **diplotricha** (*B.L.Burtt*) *I.Darbysh.* **stat. nov.** Type: Tanzania, Lushoto District, Usambara Mts, *Buchwald* 149 (K!, holo.; BM!, BR, iso.)

Leaf margin shallowly to conspicuously crenate-serrate, upper surface diplotrichous. Petioles, peduncles and pedicels with at least the shorter hairs ascending.

TANZANIA. Lushoto District: Sigi Sigoma, Dec. 1943, *Moreau* in AH 9839!; Tanga District: Kange gorge, Nov. 1956, *Milne-Redhead & Taylor* 7287! & E Usambara Mts, Marimba Forest Reserve, ± 3 km N of Kiwanda village E of Sigi River, Nov. 1986, *Iversen et al.* 86373!
DISTR. **T** 3; restricted to the coastal lowlands and foothills of the lower altitude E Usambara Mts
HAB. Rock faces, terrestrial or rarely epiphytic in lowland evergreen or semi-deciduous forest; 0–400(?–1000) m
USES. Cultivated as an ornamental, though not as widely as var. *ionantha*
CONSERVATION NOTES. This variety suffers from the same habitat threats as the typical variety in the Tanga region and is likely to have experienced significant population declines both over past decades and at present. It is also known from less than 10 localities at present. It is thus assessed as Vulnerable (VU A2c, B1ab(iii)+2ab(iii))

SYN. *S. diplotricha* B.L.Burtt in Gard. Chron., ser. 3, 122: 23 (1947) pro parte quoad type, & in Baileya 4: 163–164 (1956) & in Notes Roy. Bot. Gard. Edinb. 22: 563 (1958); Iversen in Symb. Bot. Upsal. 28: 235, t. 1B (1988); Watkins *et al.*, Wild Afr. Violet: 27 (2002)
  [*S. ionantha sensu* C.B. Clarke in F.T.A. 4(2): 500 (1906), *non* H.Wendl.]
  [*S. sp. sensu* Kolehmainen in Afr. Violet Mag. 58: 28, fig. 2 (2005) quoad spec. ex Handei village forest]

NOTE. Var. *diplotricha* superficially appears distinct from var. *ionantha* on account of the striking diplotrichous leaf indumentum, but in other respects (capsule shape and size, corolla indumentum and leaf shape and thickness) is very close to that taxon. As diplotrichy appears to have evolved on several occasions within the species it is not likely to be of great taxonomic significance, but the apparent consistency of indumentum type within populations permits the recognition of *diplotricha* at the varietal level. The type is a rather scant specimen; the fruits are atypically slender for subsp. *ionantha* but the leaf shape and indumentum are close to plants from Sigi Segoma and from an isolated population in the Handei Village Forest, E Usambaras (e.g. *Simiyu et al.* H48!), which are referable to var. *diplotricha*.

Collections from Kwamgumi in the E Usambaras vary from typical var. *diplotricha* (*Iversen* 86607! & *Borhidi et al.* 86647!) to caulescent plants with a predominantly puberulent leaf indumentum with only scattered longer hairs, the hairs appearing somewhat appressed at least on the petioles (*Frontier Tanzania* 3157! & *Simiyu et al.* H9–15!). Similar plants to the latter are also recorded from Kwamtili (e.g. *Simiyu et al.* Kwamtili 8!). These gatherings appear intermediate with subsp. *grotei*; fruiting material is required to further elucidate their identity.

b. subsp. **pendula** (*B.L.Burtt*) *I.Darbysh.* **stat. nov.** Type: Tanzania, Lushoto District, E Usambara Mts, Mt Mtai, comm. *Punter* ref. U, cult in R.B.G. Edinb., C.1686 (E!, holo.)

© The Board of Trustees of the Royal Botanic Gardens, Kew, 2006

Plants caulescent or rosulate, the stems sometimes branched in the former. Leaves pale green or tinged red-purple beneath; blade ovate to suborbicular, 3–6(–8.5) cm long, margin (crenate-)serrate, apex acute, subattenuate or more rarely obtuse, upper surface erect- to antrorse-pilose. Petioles, peduncles and pedicels spreading- to antrorse-pilose. Corolla usually deep blue-violet, margin with predominantly glandular hairs. Capsule 8–12.5 mm long, 2.3–5.5 mm diameter.

Tanzania. Lushoto District: Bulwa, Dec. 1944, *Greenway* 7042! & Mtai Forest Reserve, main ridge and summit of Mtai Peak, Nov. 1986, *Borhidi et al.* 86736! & Mtai Forest Reserve, Sept. 1996, *Kisena* 1643!

Distr. **T** 3; restricted to the northern E Usambara Mts

Hab. On rocks in moist forest; 400–1100 m

Uses. None recorded on herbarium specimens

Conservation notes. This subspecies is currently restricted to very few submontane rainforest sites in the northern East Usambara Mts, with approximately 6 sites confirmed from herbarium data. Significant forest losses over recent decades are likely to have had an impact upon populations outside protected areas, particularly at the lower end of its altitudinal range. It is therefore assessed as Vulnerable (VU B1ab(iii)+2ab(iii))

Syn. *S. pendula* B.L.Burtt in Notes Roy. Bot Gard. Edinb. 22: 561 (1958); Iversen in Symb. Bot. Upsal. 28: 243 (1988); Watkins *et al.*, Wild Afr. Violet: 40 (2002)

    S. intermedia B.L.Burtt in Notes Roy. Bot. Gard. Edinb. 25: 193 (1964); Iversen in Symb. Bot. Upsal. 28: 243 (1988); Watkins *et al.*, Wild Afr. Violet: 24 (2002). Type: Tanzania, Lushoto District, E Usambara Mts, Kigongoi, comm. *Punter* ref. V, cult in R.B.G. Edinb., C.2007 (E!, holo.), **syn. nov.**

    S. pendula B.L.Burtt var. *kizarae* B.L.Burtt in Notes Roy. Bot Gard. Edinb. 25: 193 (1964); Iversen in Symb. Bot. Upsal. 28: 243 (1988); Watkins *et al.*, Wild Afr. Violet: 41 (2002). Type: Tanzania, Lushoto District, NE Usambara Mts, Kizara, comm. *Punter* ref. X58/3589, cult. in R.B.G. Edinb., C.3066 (E!, holo.), **syn. nov.**

Note. The separation of this subspecies from subsp. *ionantha* var. *ionantha* can be difficult. In living material, the leaves are considerably less fleshy, but this character is likely phenotypic. Although plants of this subspecies sometimes develop long creeping stems, the leaves are often crowded and subrosulate in wild specimens; the trailing stems likely develop with age but plants appear able to flower when young. In cultivated material, the lateral nerves are highly conspicuous beneath, either being pale against a red-purple underside in those with anthocyanin or darker against a pale underside where this is lacking; this character can however be less conspicuous in the wild material, at least when dried.

    S. intermedia was previously separated from *S. pendula* on account of it being only "reluctantly caulescent" rather than strongly so, in the leaves being red-purple, not pale green, beneath, the hairs of the leaves being antrorse rather than erect, and the inflorescences being several-flowered, not single-flowered. Dr. Burtt however later described a variety of *S. pendula*, var. *kizarae*, which had slightly darker leaves and red-tinged petioles, 2–several-flowered inflorescences and was less strongly caulescent. This variety is clearly intermediate between *S. pendula* and *S. intermedia*. Subsequent wild collections have become available which demonstrate that stem development, degree of anthocyanin in the leaves and number of flowers per inflorescence are all variable across a continuum and are thus best considered a single entity. Var. *kizarae* was additionally separated from *S. pendula* by having more narrowly lanceolate calyx lobes. This can be discounted as there is significant overlap in this character between var. *kizarae* and plants previously attributed to both *S. pendula* and *S. intermedia*.

c. subsp. **rupicola** (*B.L.Burtt*) I.Darbysh. **stat. nov.** Type: Kenya, Kilifi District, Kaloleni, road from Mariakani to Kilifi 40 km NW of Mombasa, coll. *Bayliss*, comm. *Punter* ref. 8 of Oct. 1958, cult. in R.B.G. Edinb., C.3065 (E!, holo.)

Plants rosulate, fleshy throughout. Leaves often held partially erect particularly towards the centre of the rosette, pale green to whitish beneath; blade ovate or more rarely elliptic, 3.5–12.5 cm long, margin shallowly to coarsely crenate-serrate, apex obtuse to acute, upper surface with (sub-)erect hairs of equal length, the density and length varying between specimens. Petioles, peduncles and pedicels with (sub-)spreading hairs, the length varying between specimens. Corolla pale blue to mauve, margin with predominantly glandular hairs. Capsule (12.5–)16–30 mm long, 1.5–3(–4) mm diameter.

© The Board of Trustees of the Royal Botanic Gardens, Kew, 2006

KENYA. Kilifi District: Cha Simba, 8 km NE of Kaloleni on Kilifi Road, Feb. 1971, *Mabberley* 718!
   & Mwarakaya, 5.3 km S of turnoff on Kilifi-Kaloleni Road, May 1985, *Faden & Beentje* 85/30!
   & Kacharoroni gorge, Vitengeni River, Feb. 1988, *Robertson & Luke* 5126!
DISTR. **K** 7; not known elsewhere
HAB. Crevices in shaded or exposed limestock outcrops, or on cliffs in coastal forest; 50–250 m
USES. Cultivated as an ornamental, though not as widely as subsp. *ionantha*
CONSERVATION NOTES. This subspecies is restricted to a very few sites in coastal Kenya. The
Mwachi population is believed to be extinct (S.A. Robertson, *pers. comm.*) and there has been
continued tree cover loss and habitat degradation in the remaining sites (Eastwood *et al.* in
Bot. Mag. 15: 56 (1998)). The current populations may not regenerate due to lack of suitable
habitat (S. Simiyu, *pers. comm.*). It is therefore currently considered Endangered (EN
B1ab(iii)+2ab(iii)) but may qualify as Critically Endangered under different criteria following
more complete assessment of the populations

SYN. *S. rupicola* B.L.Burtt in Notes Roy. Bot. Gard. Edinb. 25: 193 (1964) & in Fl. Pl. Afr. 43: t.
   1720 (1976); Watkins *et al.*, Wild Afr. Violet: 43 (2002)
   [*S.* sp. nov. *sensu* Macharia *et al.* in Maesen *et al.* (eds.), Biodiv. Afr. Pl.: proceedings XIVth
     AETFAT congress: 329 (1996)]
   [*S.* "Robertson" *sensu* J. Smith in Gloxinian 48(2): 44 (1998)]
   [*S.* sp. nov. "Kacharoroni" *sensu* Eastwood et al. in Bot. Mag. 15: 56, 59 (1998); *S.*
     "Kacharoroni" *sensu* Lindqvist & Albert in K.B. 54: 366, 369 & 373 (1999) & in Syst.
     Geogr. Pl. 71: 39–41 (2001); *S.* "Kacharoni" *sensu* Watkins *et al.*, Wild Afr. Violet: 48
     (2002); *S.* sp. nov. "Kacharoni" *sensu* Schulman & Kolehmainen in Scripta Bot. Belg. 29:
     168 (2004)]
   [*S.* sp. nov. "Mwachi" *sensu* Eastwood et al. in Bot. Mag. 15: 59 (1998); *S.* "Mwache" *sensu*
     Lindqvist & Albert in K.B. 54: 366, 369 & 373 (1999) & in Syst. Geogr. Pl. 71: 40 (2001);
     S. "Mwachi" *sensu* Watkins *et al.*, Wild Afr. Violet: 50 (2002); *S.* sp. nov. "Mwachi" *sensu*
     Schulman & Kolehmainen in Scripta Bot. Belg. 29: 168 (2004)]

NOTE. Subsp. *rupicola* is unusual in growing (or at least persisting) on rather exposed limestone
outcrops where it is likely to experience periodic drought; indeed the plants are often
associated with a variety of succulents and other drought-tolerant taxa. Such plants have
developed long fleshy petioles, large thick leaves and stout fleshy stems, all of which may be
an adaptation to this environment. However, it is notable that cultivated plants derived from
these populations retain these characters. Plants collected from deeper shade, at the Mwachi
Forest Reserve (*Robertson & Luke* 6248!) and Sokoke Forest (*Campbell*, cult., R.B.G. Kew) have
smaller leaves and shorter petioles.
   Specimens from the isolated population at Kacharoroni gorge have previously been
considered a separate species, collectors having pointed principally to the large size of these
plants. The more vigorous growth habit is possibly a response to high soil nutrient content
due to the abundance of guano from the associated bat colony (S.A. Robertson, ms.).
Furthermore, similarly robust plants have been collected from the colony of *S. rupicola* at Cha
Simba to the south (*Robertson* 5461!). The Kacharoroni plants certainly have longer hairs
throughout, a character linking this population with those from the Mwachi and Sokoke
forests. These latter colonies have more elliptic leaves than typical in the subspecies.
However, these variations in leaf shape and indumentum appear insignificant when
compared to the variability recorded within the *S. ionantha* complex as a whole; they are
therefore treated here as a single entity. The plants with a pilose indumentum provide a clear
link between subsp. *rupicola* and subsp. *ionantha* var. *ionantha* Although they are usually
readily separable by the differing capsules (those of the coastal plants in the *ionantha*
complex being much shorter and more ovoid), there is a slight overlap at the extremes of the
range in fruit sizes. Plants from the Pangani Falls, Tanga District (*Simiyu et al.* H56!), appear
rather intermediate between the two taxa. Genetic data (e.g. Linqvist & Albert in Syst. Geogr.
Pl. 71: 40 (2001)) also indicates that the *rupicola* group is phylogenetically embedded within
the *ionantha* complex; it is thus reduced to subspecific status here.

   d. subsp. **grandifolia** (*B.L.Burtt*) *I.Darbysh.* **stat. nov.** Type: Tanzania, Lushoto District, W
Usambara Mts, Lutindi, comm. *Punter* ref. S, cult. in R.B.G. Edinb., C.1575 (E!, holo.)

Plants rosulate. Leaves somewhat fleshy, pale green beneath; blade ovate-elliptic to oblong-
elliptic, 4–10 cm long, margin shallowly to coarsely crenate-serrate, apex rounded to obtuse, upper
surface erect-pilose. Petioles, peduncles and pedicels spreading-pilose. Corolla deep blue-violet,
margin with predominantly glandular hairs. Capsule 10–18 mm long, 2.5–3.8 mm diameter.

© The Board of Trustees of the Royal Botanic Gardens, Kew, 2006

TANZANIA. Lushoto District: Lutindi, near Lewa village, Dec. 1991, *Raistrick* SP/6! & Massange Forest, Jan. 1992, *Raistrick* SP/7! & between Matundsi and Mashindei peaks SW of Ambangulu Tea Estate, Feb. 1985, *Borhidi et al.* 85463!

DISTR. **T** 3; restricted to the southern W Usambara Mts

HAB. Shaded wet rockfaces, banks and streamside rocks in rainforest; 750–1400 m

USES. Cultivated as an ornamental, though not as widely as subsp. *ionantha*

CONSERVATION NOTES. This subspecies is currently known from a highly restricted range within the W Usambara Mts, being recorded from approximately four sites. Raistrick (SP/6) noted that very few plants remained in the population near Lewa village due to agricultural encroachment. It is thus assessed as Endangered (EN B1ab(iii)+2ab(iii))

SYN. *S. grandifolia* B.L.Burtt in Notes Roy. Bot. Gard. Edinb. 22: 560 (1958); Iversen in Symb. Bot. Upsal. 28: 241 (1988); Watkins *et al.*, Wild Afr. Violet: 29 (2002)

NOTE. This subspecies is very close to subsp. *ionantha*, but the consistent combination of pale undersides to the leaves, dark (blue-)violet flowers and rather elongate fruits allow its separation. In subsp. *ionantha*, the leaf underside and flower colour appear more closely linked, those with little anthocyanin in the leaves tending to have pale flowers. The habitat also differs between the two taxa, subsp. *grandifolia* being the submontane, wet forest equivalent of the coastal, semi-deciduous forest taxon subsp. *ionantha*. The less fleshy leaves of subsp. *grandifolia* in comparison to subsp. *ionantha* is thus likely to be a phenotypic response.

e. subsp. **grotei** (*Engl.*) *I.Darbysh.* **stat. nov.** Type: Tanzania, Lushoto District, E Usambara Mts, near Amani, *Grote* 3708 (B†, holo., K!, illus.; EA!, iso.)

Plants rosulate or caulescent and trailing, internodes sparsely pilose. Leaves pale green or tinged purple beneath; blade broadly ovate, suborbicular or (oblong-)elliptic, 3–10 cm long, margin coarsely to obscurely (crenate-)serrate, apex obtuse, rounded or more rarely acute to subattenuate, upper surface with appressed hairs of subequal or more often variable length, at least some short, sometimes diplotrichous, the longest hairs sometimes arced or suberect. Petioles, peduncles and pedicels with predominantly appressed or strongly ascending hairs, the longest hairs sometimes subspreading. Inflorescences 1–4(–6)-flowered. Corolla pale mauve to deep blue-violet, margin with mixed glandular and eglandular or predominantly eglandular hairs. Capsule 8–20 mm long, 1.5–3 mm diameter.

TANZANIA. Lushoto District: E Usambara Mts, Kwamkoro to Kihuhwi, Dec. 1936, *Greenway* 4797! & Kwamkoro Forest Reserve, from Kwamkoro Tea Estate to Kimbo summit, Oct. 1986, *Iversen et al.* 86201!; Tanga District: E Usambara Mts, Mt Mlinga, Dec. 1991, *Raistrick* SP/3!

DISTR. **T** 3; restricted to the central and southern E Usambara Mts

HAB. Rock faces and terrestrial in moist forest; 900–1100 m

USES. None recorded on herbarium specimens

CONSERVATION NOTES. This taxon has been recorded from many of the forests of the central and southern E Usambara Mts and healthy populations remain at some of the protected sites. However, its range is centred around the town of Amani, around which large areas of forest have been cleared for forestry and agriculture, thus many previous sites for subsp. *grotei* have been lost. It is therefore provisionally assessed as Vulnerable (VU A2c)

SYN. *S. grotei* Engl. in E.J. 57: 202 (1921); B.L. Burtt in Notes Roy. Bot. Gard. Edinb. 22: 566 (1958); Iversen in Symb. Bot. Upsal. 28: 243 (1988); Watkins *et al.*, Wild Afr. Violet: 30 (2002)

S. magungensis E.P.Roberts in Afr. Violet Mag. 3(4): 6 (1950); B.L. Burtt in Notes Roy. Bot. Gard. Edinb. 22: 567 (1958) & in Notes Roy. Bot. Gard. Edinb. 25: 195 (1964); Iversen in Symb. Bot. Upsal. 28: 243 (1988); Watkins *et al.*, Wild Afr. Violet: 34 (2002). Type: Tanzania, Lushoto District, E Usambara Mts, Magunga, cult. in Michigan State Coll., *E.P. Roberts* 1 (Michigan holo.), **syn. nov.**

S. amaniensis E.P.Roberts in Afr. Violet Mag. 4(2): 7 (1950); B.L. Burtt in Notes Roy. Bot. Gard. Edinb. 22: 566 (1958). Type: Tanzania, Lushoto District, E Usambara Mts, near Amani, cult. Michigan State Coll., *E.P. Roberts* 2 (Michigan, holo.) **syn. nov.**

S. confusa B.L.Burtt in Baileya 4: 164 (1956) & in Notes Roy. Bot. Gard. Edinb. 22: 566 (1958); Iversen in Symb. Bot. Upsal. 28: 243 (1988); Watkins *et al.*, Wild Afr. Violet: 25 (2002). Type: Tanzania, Lushoto District, E Usambara Mts, cult. in R.B.G. Kew (E!, holo.), **syn. nov.**

© The Board of Trustees of the Royal Botanic Gardens, Kew, 2006

S. *difficilis* B.L.Burtt in Notes Roy. Bot. Gard. Edinb. 22: 565 (1958); Iversen in Symb. Bot. Upsal. 28: 243 (1988); Watkins *et al.*, Wild Afr. Violet: 26 (2002). Type: Tanzania, Lushoto District, E Usambara Mts, Sigi headwaters above Monga, comm. *Punter* ref. Q, cult. in R.B.G. Edinb., C.1578 (E!, holo.), **syn. nov.**

S. *magungensis* E.P.Roberts var. *minima* B.L.Burtt in Notes Roy. Bot. Gard. Edinb. 25: 195 (1964); Iversen in Symb. Bot. Upsal. 28: 243 (1988); Watkins *et al.*, Wild Afr. Violet: 35 (2002). Type: Tanzania, Lushoto District, Mavoera estate near Amani, comm. *Punter* ref. 59/4352, cult. in R.B.G. Edinb., C.3724 (E!, holo.), **syn. nov.**

[S. *diplotricha sensu* B.L. Burtt in Gard. Chron., ser. 3, 122: 23 (1947) pro parte quoad spec. cult. R.B.G. Kew]

Note. The distinction between S. *grotei* and S. *magungensis* (within which S. *amaniensis* was reduced to synonymy by B.L. Burtt in Notes Roy. Bot. Gard. Edinb. 25: 195 (1964)) has always been rather unclear. The name S. *grotei* has usually been applied to larger specimens with long petioles and leaves with distinctly crenate margins. Burtt also indicated that S. *grotei* has paler flowers than S. *magungensis*. Both entities have been recorded growing together on Mt Mlinga by both Greenway and Raistrick. However, analysis of the Mlinga material indicates that there is considerable variability in flower colour, colour of the underside of the leaf, and size of the plants there; separation into two discrete taxa does not seem possible. A third entity recorded on Mt Mlinga, with a rosulate habit, is typical of plants previously assigned to S. *confusa*. Indeed, the differing habit appears to be the only significant separating character. Whilst both the caulescent and rosulate forms do develop under similar conditions in cultivation, thus not being entirely phenotypic or age-related, the recognition of the two forms in the wild is highly problematic. Plants of *Saintpaulia* often flower when young, thus plants which later become caulescent may be rosulate at the time of observation. More significantly, plants displaying an intermediate growth form which is subrosulate or with short internodes developing are recorded, for example from the Magrotto area (e.g. *Mwasumbi & Suleimani* 17450!). The rosulate plants tend to have more flowers per inflorescence ((1–)2–4(–6)) than the caulescent plants (1–2(–3)) but there is considerable overlap. It is here considered that the three entities all represent a single taxon. Mt Mlinga therefore displays a good example of the range of variation possible within a *Saintpaulia* population.

Plants previously named S. *difficilis* have more narrowly ovate to ovate-elliptic leaves with an acute apex and rather coarsely serrate margins, and the indumentum is more clearly diplotrichous, with long arced to suberect hairs and short appressed hairs. Such plants appear rather intermediate between subsp. *grotei* and subsp. *ionantha* var. *diplotricha* and indeed may form a continuum between the two taxa. However, they are placed within subsp. *grotei* here on account of their having more elongate, narrower fruit (length to width ratio (2.5–)5–8.8:1) and less numerous glandular hairs on the margin of the corolla than in subsp. *ionantha*, in addition to the fact that they have an appressed element to their leaf indumentum. These populations are perhaps worthy of varietal status, though specimens intermediate with typical subsp. *grotei* are regularly recorded (e.g. *Simiyu et al.* H78! from Ngua Forest Reserve, and *Raistrick* SP/1! from Kiganga Forest Reserve).

S. *magungensis* var. *minima* was described on the basis of two cultivated plants from material from a single locality, it being a more slender plant than S. "*magungensis*" with smaller leaves and flowers. In view of the wide range in flower and leaf size recorded in most *Saintpaulia* taxa, this appears a rather insignificant difference. Furthermore, plants grown at R.B.G., Kew from the type plant produced considerably larger leaves and flowers, well within the range typical of subsp. *grotei*. A wild collection from the type locality (*Iversen & Steiner* 86683!) also has similarly larger leaves and flowers.

f. subsp. **orbicularis** (*B.L.Burtt*) *I.Darbysh.* **stat. nov.** Type: Tanzania, Lushoto District, W Usambara Mts, Sakare, Ambangulu, *Moreau* 2 (K!, holo.)

Plants rosulate. Leaves pale green beneath; blade orbicular to ovate, 3–7.5 cm long, margin shallowly crenate-serrate, apex obtuse or rarely acute, upper surface with short, strictly appressed hairs of somewhat variable length. Petioles, peduncles and pedicels with predominantly ascending or subappressed hairs. Inflorescences (2–)4–12-flowered. Corolla white to pale mauve, always with a violet mouth and base to the upper lip, rarely violet throughout, margin with predominantly glandular hairs. Capsule 11–25.5 mm long, 1.4–3 mm diameter.

Tanzania. Lushoto District: W Usambara Mts, Ambangulu, Jan. 1971, *Archbold* 1347! & 8 km NW of Ambangulu Tea Estate along the road to Lutindi, Feb. 1985, *Borhidi et al.* 85488! & Ambangulu Tea Estate, 8 km along road to Lutindi, Dec. 1991, *Raistrick* SP/5!

© The Board of Trustees of the Royal Botanic Gardens, Kew, 2006

DISTR. **T** 3; restricted to the southern W Usambara Mts

HAB. Wet rocks by streams and waterfalls; 900–1250 m

USES. None recorded on herbarium specimens

CONSERVATION NOTES. This taxon is resticted to the Ambangulu area of the W Usambara Mts, with several subpopulations known. Raistrick (SP/5) noted some healthy plants growing at two sites. Although significant areas of forest remain in this area, plantations of cardamom are encroaching into some forests and likely threaten the native flora (Iversen, SAREC Usambara rain for. proj. report: 11 (1988)). Subsp. *orbicularis* is thus assessed as Endangered (EN B1ab(iii)+2ab(iii))

SYN. *S. orbicularis* B.L.Burtt in Gard. Chron., ser. 3, 122: 23 (1947) & in Notes Roy. Bot. Gard. Edinb. 22: 564 (1958); Iversen in Symb. Bot. Upsal. 28: 242 (1988); Watkins *et al.*, Wild Afr. Violet: 38 (2002)

    S. orbicularis B.L.Burtt var. *purpurea* B.L.Burtt in Notes Roy. Bot. Gard. Edinb. 25: 194 (1964) & in Bot. Mag. 181, t. 722 (1976); Iversen in Symb. Bot. Upsal. 28: 242 (1988); Watkins *et al.*, Wild Afr. Violet: 39 (2002). Type: Tanzania, Lushoto District, W Usambara Mts, Ambangulu, comm. *Punter* ref. 3 of Oct. 1958, cult. in R.B.G. Edinb., C.2959 (E!, holo.) **syn. nov.**

NOTE. This subspecies is the west Usambaran equivalent of subsp. *grotei*. The leaf shape and indumentum are very close to that subspecies, but it differs in having longer and more slender capsules, generally more numerous flowers per inflorescence (though with considerable overlap) and predominantly glandular hairs on the corolla margin. The anthers are rather small in this subspecies, being 1.4–1.8(–2) mm long. Both this character and the most common corolla colour (pale with a dark "eye") place this subspecies rather close to *S. shumensis*.

    Var. *purpurea* was recognised on account of having uniformly violet flowers with more oblong upper lobes. In view of the large variation seen in several other taxa of *Saintpaulia*, the difference in flower colour is not considered of great significance. The lobes of the type specimen do appear more oblong, however the plants grown at Kew from the Edinburgh stock include well-developed flowers with more rounded lobes. Wild specimens of the all-violet flowered form are not known to the author.

    g. subsp. **velutina** (*B.L.Burtt*) *I.Darbysh.* **stat. nov.** Type: Tanzania, Lushoto District, W Usambara Mts, Balangai, 8 km from Sakarre, comm. *Punter* ref. D, cult. in R.B.G. Edinb., C.1579 (E!, holo.)

Plants rosulate. Leaves strongly purple or more rarely green beneath; blade broadly ovate to orbicular, 2–8.5 cm long, margin subentire, crenate or crenate-serrate, apex rounded to broadly obtuse, upper surface densely erect puberulent, with or without interspersed long erect to arced hairs. Petioles, peduncles and pedicels with spreading, ascending or suberect hairs of variable length. Inflorescences 2–6-flowered. Corolla deep blue-purple throughout or with the lobes paler towards the margins, rarely paler throughout, margin variously with predominantly glandular or predominantly eglandular hairs or with a mixture of both. Capsule 11–20 mm long, 2–3 mm diameter.

TANZANIA. Lushoto District: Balangai, Jan. 1953, *Faulkner* 1126!; Morogoro District: Kanga Mt, July 1983, *Polhill & Lovett* 4956! & ibid., Dec. 1987, *Lovett & D.W. Thomas* 2787!

DISTR. **T** 3, 6; restricted to the W Usambara, Nguru and Uluguru Mts

HAB. On rocks, lower tree trunks, damp banks and by streams in forest; (350–)700–1500 m

USES. None recorded on herbarium specimens

CONSERVATION NOTES. This taxon is currently known from less than 10 sites over a rather broad distribution in the Tanzanian highlands. Populations at Kanga in the Nguru Mts appear healthy and largely unthreatened. However, forest at the type locality, Balangai in the W Usambaras, has been seriously degraded in recent decades through timber extraction and agricultural encroachment. It is therefore assessed as Vulnerable (VU B2ab(iii))

SYN. *S. velutina* B.L.Burtt in Notes Roy. Bot. Gard. Edinb. 22: 563 (1958) & in Notes Roy. Bot. Gard. Edinb. 25: 194 (1964); Iversen in Symb. Bot. Upsal. 28: 242 (1988); Watkins *et al.*, Wild Afr. Violet: 47 (2002)

    S. brevipilosa B.L.Burtt in Notes Roy. Bot. Gard. Edinb. 25: 193 (1964); Watkins *et al.*, Wild Afr. Violet: 47 (2002). Type: Tanzania, Morogoro District, Nguru Mts, Lulaga, comm. *Punter* ref. 59/4350, cult in R.B.G. Edinb., C.3827 (E!, holo.), **syn. nov.**

© The Board of Trustees of the Royal Botanic Gardens, Kew, 2006

NOTE. The type collections of *S. velutina* and *S. brevipilosa* (and the resultant cultivated stock) appear very different and it is with reluctance that *S. brevipilosa* is here reduced to synonymy. However, a review of the material from Kanga Mt (**T** 6), clearly demonstrates the variability possible within this taxon. Plants from the lowest altitudes (*Polhill & Lovett* 4956!, 750 m alt.) have large, subentire leaves with a puberulent indumentum (although close examination reveals the presence of somewhat longer hairs); such plants are close to the type of *S. brevipilosa*. A specimen from 1000–1100 m alt. (*Pócs* 6137/C!) has a more variable indumentum, with some of the younger leaves clearly having some longer hairs. At higher elevations (1200–1500 m alt., e.g. *Lovett & D.W. Thomas* 2787!), the leaves are smaller and clearly crenate and have conspicuous long hairs interspersed within the puberulent indumentum. Such specimens are practically inseparable from plants from Balangai in the Usambara Mts, the type location of subsp. *velutina*. Slight differences do occur, notably that in the Kanga material the marginal hairs of the corolla are predominantly eglandular (glandular in Balangai), the petioles, peduncles and pedicels have subappressed to ascending hairs (spreading in Balangai) and the leaf margin is subentire to crenate (crenate-serrate in Balangai). However, such differences are inconsistent; for example, in the type specimen of *S. brevipilosa* the hairs on the petioles, peduncles and pedicels are spreading and the marginal hairs of the corolla are mixed glandular and eglandular. The fruits of the Usambara populations are currently known only from a single, immature specimen (*Iversen et al.* 84263!); and appear somewhat shorter than the Nguru plants, though some are cylindric. It is the narrower cylindric fruits that principally separate this taxon from subsp. *ionantha* var. *diplotricha*, although the more dense minute hairs in subsp. *velutina* give the leaves a more velvety texture. The habitat also differs significantly between the two taxa.

A collection from 720 m alt. in the Mkungwe Catchment Forest Reserve in the northeast Uluguru Mts (*Jannerup & Mhoro* 271!) with diplotrichous leaves which are subentire when mature is placed within this taxon. This represents the only collection of the *S. ionantha* complex known from the Uluguru Range. It is interesting in having scattered long glandular hairs on some of the pedicels. However, some plants lack this character, thus it is not considered sufficiently consistent to be of taxonomic significance. Further analysis of this population, including collection of fruits, is desirable to confirm its placement.

The taxon recorded as *S.* cf. *velutina* B.L.Burtt by Lindqvist & Albert (in K.B. 54: 367 (1999); = *S.* sp. nov. Mhonda *sensu* Schulman & Kolehmainen in Scripta Bot. Belg. 29: 168 (2004)) possibly refers to this subspecies. The voucher (not seen in the current study) is recorded as having been collected from the Dibohelo River near Mhonda Mission in the Nguru Mts. The genetic data suggests that the Nguru material (both *S. brevipilosa* and *S.* cf. *velutina*) is genetically distinct from the Usambara group (which includes the type locality of *S. velutina*) which disagrees with the current circumscription of subsp. *velutina*.

i. subsp. **nitida** (*B.L.Burtt*) *I.Darbysh.* **stat. nov.** Type: Tanzania, Morogoro District, Nguru Mts, near Morogoro, comm. *Cox* ref. A, cult. in R.B.G. Edinb., C.1557 (E!, holo.)

Plants rosulate. Leaves pale green to purplish beneath; blade broadly ovate to suborbicular, 2–7.5 cm long, margin subentire to shallowly crenate(-serrate), apex rounded or obtuse, upper surface densely appressed-pubescent, the hairs short and rather even. Petioles, peduncles and pedicels with (sub)appressed hairs. Inflorescences 1–8-flowered. Corolla deep blue-purple, margin with predominantly eglandular hairs. Capsule 16–22 mm long, 2–2.5 mm diameter.

TANZANIA. Morogoro District: Nguru Mts, Mkobwe, NW side, near Turiani, Mar. 1953, *Drummond & Hemsley* 1862! & [Koluhamba] Ruhamba, near Turiani, Dec. 1953, *Eggeling* 6776! & Nguru S Forest Reserve, head of valley behind Mhonda Mission, Feb. 1971, *Mabberley & Pócs* 695!
DISTR. **T** 6; restricted to the Nguru Mts
HAB. Rocks in forest and by forest streams; 800–1200(–1500) m
USES. None recorded on herbarium specimens
CONSERVATION NOTES. This taxon has been noted as locally common both near Mhonda Mission (*Mabberley & Pócs* 695) and at Ruhamba (*Semsei* 1463) in the Nguru Mts, and may not be threatened here. Mid-altitude forests in these mountains are however threatened by human encroachment, with some areas cleared or degraded. The extent to which this impacts upon subsp. *nitida* is unclear; it is therefore assesed as Data Deficient (DD)

SYN. *S. nitida* B.L.Burtt in Notes Roy. Bot. Gard. Edinb. 22: 564 (1958); Watkins *et al.*, Wild Afr. Violet: 37 (2002)
  [*S. pusilla sensu* Lindqvist & Albert in K.B. 54: 371 (1999) pro parte quoad *Congdon* 205, non Engl.]

© The Board of Trustees of the Royal Botanic Gardens, Kew, 2006

NOTE. This subspecies is the southern Nguru Mts equivalent of subsp. *velutina*, sharing a similar leaf shape and margin, corolla indumentum and capsule shape to the lower altitude Nguru populations of that taxon. They are distinguishable by the very different vegetative indumentum and the thin, less fleshy, shiny leaves. One collection from Kanga Mt (*Haston* 77!) however appears intermediate between the two, having dense subapressed short hairs and suberect long hairs on the leaves. Subsp. *nitida* can also appear very similar to densely pubescent variants of subsp. *grotei* although the capsules are generally longer. It is also close to subsp. *occidentalis*, particularly in capsule shape; the latter however has a less dense indumentum and a caulescent habit.

g. subsp. **mafiensis** *I.Darbysh.* & *Pócs* **subsp. nov.** ab omnibus ceteris subspeciebus pagina superiore foliorum pilos sparsissimos breves ferenti differt. Type: Tanzania, Lushoto District, Mafi Hill, *Mziray* & *Temu* 85286 (UPS!, holo.; E!, iso.)

Plants rosulate. Leaves purplish beneath; blade suborbicular, 3.5–5 cm long, margin shallowly crenate, apex rounded, upper surface with very sparse short hairs, these possibly appressed. Petioles, peduncles and pedicels with antrorse or ascending long hairs and shorter subapressed hairs. Inflorescences 1–4-flowered. Corolla violet, margin with predominantly eglandular hairs. Capsule 14–24 mm long, 1.7–2.3 mm diameter.

TANZANIA. Lushoto District: Mafi Hill, Kwakulonge stream, Jan. 1985, *Mziray* & *Temu* 85286! (type)
DISTR. **T** 3; restricted to the type location
HAB. Shaded cliffs in submontane moist forest; 1300 m
USES. None recorded on herbarium specimens
CONSERVATION NOTES. This taxon is restricted to a single location in the isolated rainforests of Mafi Hill to the south of the Usambara range. The collecting notes of the single specimen give no indication of the abundance at that site. The submontane rainforest habitat here was however recorded as largely undisturbed and extensive in the 1980s (Iversen in Symb. Bot. Upsal. 28: 242 (1988)), thus the status of Critically Endangered recorded by Eastwood *et al.* (in Bot. Mag. 15: 59 1998) does not appear applicable. However, with such a limited range, this subspecies is susceptible to serious decline through future stochastic events, such as rockfalls or prolonged local drought, or through future human activity at this site; it is therefore considered Vulnerable (VU D2)

SYN. [*S.* "*mafiensis*" *sensu* Pócs in Iversen, SAREC Usambara rain for. proj. report: 24, 27, 35 (1988) & in Iversen in Symb. Bot. Upsal. 28: 242, t. 1A (1988); Baatvik in Fragm. Flor. Geobot. Suppl. 2, 1: 105, 108 (1993); Lindqvist & Albert in Syst. Geogr. Pl. 71: 38 (2001); *S.* sp. nov. "*mafiensis*" *sensu* Eastwood et al. in Bot. Mag. 15: 59 (1998)]
[*S.* "Mafia Hills" *sensu* Watkins *et al.*, Wild Afr. Violet: 49 (2002)]

NOTE. This taxon has remained undescribed for over 20 years despite the belief by several researchers that it represented a new species. Dr. Burtt (E, ms.) suggested that it was similar to *S. nitida* (= *S. ionantha* subsp. *nitida*), presumably on the basis of the narrowly cylindric fruits and the suborbicular leaves with a shallowly crenate margin. It also appears close to some populations of subsp. *velutina*. However, the leaf indumentum is characteristic and it is on this basis that this isolated subspecies is recognised.

f. subsp. **occidentalis** (*B.L.Burtt*) *I.Darbysh.* **comb. & stat. nov.** Type: Tanzania, Lushoto District, W Usambara Mts, Mlalo, Bagai, comm. *Punter* ref. 59/4355, cult. in R.B.G. Edinb., C.3854 (E!, holo.)

Plants caulescent, trailing, internodes rather densely appressed-pubescent. Leaves dark green and shining above, purplish beneath; blade ovate to elliptic, 3–6.5 cm long, margin shallowly crenate-serrate, apex subattenuate, upper surface with short appressed hairs of rather equal length. Petioles, peduncles and pedicels with predominantly subappressed hairs. Inflorescences (1–)2–4-flowered. Corolla mauve-blue, margin with predominantly eglandular hairs. Capsule 18–25 mm long, 2.2–3 mm diameter.

TANZANIA. Lushoto District: Mlalo, Bagai, Sept. 1959, *Punter* SD 3005/1! & Bagai, Shagayu Forest, Dec. 1991, *Raistrick* SP/4!
DISTR. **T** 3; restricted to the northern W Usambara Mts
HAB. Streamside banks in moist forest; 1400 m

© The Board of Trustees of the Royal Botanic Gardens, Kew, 2006

USES. None recorded on herbarium specimens

CONSERVATION NOTES. This subspecies is apparently known from only one location at present, where a large and healthy colony is recorded. This area still holds extensive natural forest wehich is now protected as a forest reserve, but illegal pit-sawing has been recorded (Iversen, SAREC Usambara rain for. proj. report: 12 (1988)). The single population must therefore be considered threatened and subsp. *occidentalis* is assessed as Critically Endangered (CR B1ab(iii)+2ab(iii))

SYN. *S. magungensis* E.P.Roberts var. *occidentalis* B.L.Burtt in Notes Roy. Bot. Gard. Edinb. 25: 195 (1964); Iversen in Symb. Bot. Upsal. 28: 242 (1988); Watkins *et al.*, Wild Afr. Violet: 36 (2002)

NOTE. This highly isolated population was previously placed as a variety of *S. magungensis* (= *S. ionantha* subsp. *grotei*) by Dr. Burtt but its combination of rather densely hairy stems, uniformly hairy leaves, long calyx lobes and long slender fruits make it the most distinct of the subspecies with caulescent forms. In terms of indumentum and leaf lustre, it is rather close to *S. teitensis*; an evolutionary link between these two taxa is not unlikely as subsp. *occidentalis* is geographically the closest taxon of the *S. ionantha* complex to *S. teitensis*. S. Simiyu (*pers. comm.*) suggests that subsp. *occidentalis* can also develop subpeltate leaves as in *S. teitensis*, but this has not yet been recorded in herbarium material or cultivated stock. Further study of this taxon is required to confirm its relationship to both *S. ionantha* and *S. teitensis*.

6. **S. teitensis** *B.L.Burtt* in Notes Roy. Bot. Gard. Edinb. 22: 559 (1958) & in Notes Roy. Bot. Gard. Edinb. 25: 192 (1964); Watkins *et al.*, Wild Afr. Violet: 45 (2002). Type: Kenya, Teita District, Mbololo Hill, *Boy Joanna* in CM 8982 (K sheet 2!, lecto., selected here; EA!, K sheet 3!, isolecto.)

Robust rosulate herb. Stems stout and fleshy below the current leaf cluster, to 12 cm long, pilose when young. Leaves rather dark green and shining above, pale green to purplish beneath; blade variable, broadly ovate, (oblong-)elliptic, suborbicular or somewhat obovate, 4–8.5 cm long, 3–7 cm wide, base shallowly cordate when young, becoming subpeltate with maturity, margin shallowly serrate, apex broadly obtuse, upper surface rather densely pilose, the hairs curved or subappressed, dense on the margins; lateral nerves 4–5 pairs; petiole 4–23 cm long, densely (sub-)appressed-pubescent. Inflorescences 3–4-flowered; peduncles 3–7 cm long; pedicels 11–18 mm long, both densely (sub-)appressed-pubescent; bracts linear, 2.5–5.5 mm long, appressed-pubescent. Calyx lobes (linear-)lanceolate, 2.5–4.5 mm long in flower, extending to 4–5.5 mm in fruit, outer surface densely appressed-pubescent. Corolla purple, pubescent on the limb and apex of the tube outside, the hairs reducing in length towards the apex of the lobes, with short, predominantly glandular hairs along the lobe margins; tube 2–3 mm long; limb 22–31 mm long; upper lip 8–12 mm long, lobes (obovate-)rounded, 4–9.5 mm long, 3.5–9.5 mm wide; lower lip 11–16.5 mm long, lobes obovate-rounded, 8.5–13.5 mm long, 6–13.5 mm wide. Filaments flattened, 3–3.5 mm long; anther thecae 2–2.5 mm long; staminodes 2, to 1.5 mm long. Ovary conical, 3–4 mm long, densely appressed-pubescent; style 5.5–7.5 mm long; stigma shallowly bilobed, 0.4–0.5 mm diameter, papillose. Capsule cylindric, 17–23 mm long, 2–3.5 mm diameter, apex tapered, shortly pubescent, style ± persistent. Seeds 0.5–0.55 mm long, verruculose with longitudinally ridges.

KENYA. Teita District: Mbololo Hill, Sept.–Oct. 1938, *Boy Joanna* in CM 8982! (type) & ibid., June 1985, *Taita Hills Expedition* 1170! & ibid., Feb. 1996, *Simiyu* 174!

DISTR. **K 7**; restricted to the Taita Hills

HAB. In rock crevices and on cliff faces in forest, more rarely epiphytic or terrestrial; 1400–1850 m

USES. None recorded on herbarium specimens

CONSERVATION NOTES. This species appears restricted to a single site in the wild, Mbololo Hill, where it is estimated to occupy an area of less than 1 km² with less than 2500 individual plants (Eastwood & Maunder, Cons. Assess. *Saintpaulia* taxa cult. within R.B.G. Kew: 16 (1995)). It was recorded as locally common there in the 1980s (*Taita Hills Expedition* 1170), and five

© The Board of Trustees of the Royal Botanic Gardens, Kew, 2006

subpopulations are believed to be extant (S. Simiyu, *pers. comm.*). Forest is, however, severely threatened by human exploitation in the Taita Hills even at higher elevations. It is therefore considered to be Critically Endangered (CR B1ab(iii)+2ab(iii))

NOTE. This species is clearly closely allied to the *S. ionantha* complex, the flowers being largely identical. The large rosettes of leaves with long petioles, and the fusion of the amplexicaul lobes of the leaf base to form peltate mature leaves are however distinctive (but see note to *S. ionantha* subsp. *occidentalis*). The latter character can be difficult to observe in herbarium material. The inflorescence is often shorter than the leaves, but this character is sometimes seen in other large rosulate taxa such as *S. ionantha* subsp. *ionantha* and subsp. *rupicola*.

A holotype was not designated in the protologue of *S. teitensis*; the K sheet labelled "Sheet II" of *Boy Joanna* in Coryndon Museum 8982 is therefore selected here as a lectotype as this has both flowers and fruits; K sheet I is currently missing.

© The Board of Trustees of the Royal Botanic Gardens, Kew, 2006

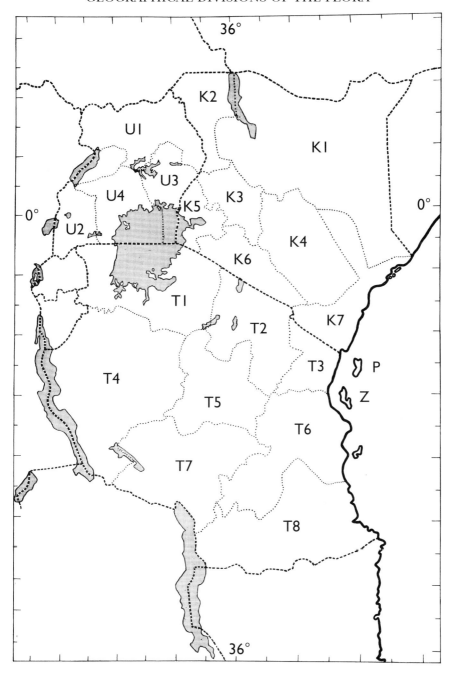

© The Board of Trustees of the Royal Botanic Gardens, Kew, 2006

INDEX TO GESNERIACEAE

© The Board of Trustees of the Royal Botanic Gardens, Kew, 2006

New names validated in this part

Saintpaulia ionantha *H.Wendl.* subsp. **grandifolia** (*B.L.Burtt*) *I.Darbysh.* **stat. nov.**
Saintpaulia ionantha *H.Wendl.* subsp. **grotei** (*Engl.*) *I.Darbysh.* **stat. nov.**
Saintpaulia ionantha *H.Wendl.* subsp. **ionantha** var. **diplotricha** (*B.L.Burtt*) *I.Darbysh.* **stat. nov.**
Saintpaulia ionantha *H.Wendl.* subsp. **mafiensis** *I.Darbysh. & Pócs* **subsp. nov.**
Saintpaulia ionantha *H.Wendl.* subsp. **nitida** (*B.L.Burtt*) *I.Darbysh.* **stat. nov.**
Saintpaulia ionantha *H.Wendl.* subsp. **occidentalis** (*B.L.Burtt*) *I.Darbysh.* **comb. & stat. nov.**
Saintpaulia ionantha *H.Wendl.* subsp. **orbicularis** (*B.L.Burtt*) *I.Darbysh.* **stat. nov.**
Saintpaulia ionantha *H.Wendl.* subsp. **pendula** (*B.L.Burtt*) *I.Darbysh.* **stat. nov.**
Saintpaulia ionantha *H.Wendl.* subsp. **rupicola** (*B.L.Burtt*) *I.Darbysh.* **stat. nov.**
Saintpaulia ionantha *H.Wendl.* subsp. **velutina** (*B.L.Burtt*) *I.Darbysh.* **stat. nov.**
Streptocarpus **albus** (*E.A.Bruce*) *I.Darbysh.* **comb. nov.**
Streptocarpus **albus** (*E.A.Bruce*) *I.Darbysh.* subsp. **edwardsii** (*Weigend*) *I.Darbysh.* **comb. nov.**
Streptocarpus **heckmannianus** (*Engl.*) *I.Darbysh.* **comb. nov.**
Streptocarpus **heckmannianus** (*Engl.*) *I.Darbysh.* subsp. **gracilis** (*E.A.Bruce*) *I.Darbysh.* **comb. nov.**
Streptocarpus **subscandens** (*B.L.Burtt*) *I.Darbysh.* **comb. nov.**
Streptocarpus **eylesii** *S.Moore* subsp. **chalensis** *I.Darbysh.* **subsp. nov.**
Streptocarpus **mbeyensis** *I.Darbysh.* **sp. nov.**
Streptocarpus **rhodesianus** *S.Moore* subsp. **grandiflorus** *I.Darbysh.* **subsp. nov.**

© The Board of Trustees of the Royal Botanic Gardens, Kew, 2006

PLANTS PEOPLE
POSSIBILITIES

© The Board of Trustees of the Royal Botanic Gardens, Kew 2006

Illustrations copyright © Contributing artists

The authors and illustrators have asserted their rights to be identified as the authors of this
work in accordance with the Copyright, Designs and Patents Act 1988.

All rights reserved. No part of this publication may be reproduced, stored in a retrieval
system, or transmitted, in any form, or by any means, electronic, mechanical, photocopying,
recording or otherwise, without written permission of the publisher unless in accordance with
the provisions of the Copyright Designs and Patents Act 1988.

Great care has been taken to maintain the accuracy of the information contained in this
work. However, neither the publisher, the editors nor authors can be held responsible for any
consequences arising from use of the information contained herein.

First published in 2006 by
Royal Botanic Gardens, Kew
Richmond, Surrey, TW9 3AB, UK
www.kew.org

ISBN 1 84246 166 4

British Library Cataloguing in Publication Data
A catalogue record for this book is available from the British Library

Design and typesetting by Margaret Newman,
Kew Publishing, Royal Botanic Gardens, Kew.

Printed in the UK by Hobbs the Printers

For information or to purchase all Kew titles please visit
www.kewbooks.com or email publishing@kew.org

All proceeds go to support Kew's work in saving the world's plants for life